Society of Earth Scientists Series

Series editor

Satish C. Tripathi, Lucknow, India

The Society of Earth Scientists Series aims to publish selected conference proceedings, monographs, edited topical books/text books by leading scientists and experts in the field of geophysics, geology, atmospheric and environmental science, meteorology and oceanography as Special Publications of The Society of Earth Scientists. The objective is to highlight recent multidisciplinary scientific research and to strengthen the scientific literature related to Earth Sciences. Quality scientific contributions from all across the Globe are invited for publication under this series.

More information about this series at http://www.springer.com/series/8785

Rajesh Joshi · Kireet Kumar · Lok Man S. Palni
Editors

Dynamics of Climate Change and Water Resources of Northwestern Himalaya

Springer

Editors
Rajesh Joshi
Watershed Processes and Management
G.B. Pant Institute of Himalayan
Environment and Development
Almora
Uttarakhand
India

Lok Man S. Palni
Graphic Era University
Dehradun
Uttarakhand
India

Kireet Kumar
Watershed Processes and Management
G.B. Pant Institute of Himalayan
Environment and Development
Almora
Uttarakhand
India

ISSN 2194-9204 ISSN 2194-9212 (electronic)
Society of Earth Scientists Series
ISBN 978-3-319-13742-1 ISBN 978-3-319-13743-8 (eBook)
DOI 10.1007/978-3-319-13743-8

Library of Congress Control Number: 2014956214

Springer Cham Heidelberg New York Dordrecht London

Printed on acid-free paper

Springer International Publishing AG Switzerland is part of Springer Science+Business Media (www.springer.com)

From the Institute

G.B. Pant Institute of Himalayan Environment and Development (GBPIHED), founded in 1988 during the birth centenary year of Bharat Ratna Pt. Govind Ballabh Pant, is an autonomous Institute of Ministry of Environment, Forests & Climate Change (MoEF & CC), Government of India. The Institute is a focal agency with the mandate to advance scientific knowledge, evolve integrated management strategies, demonstrate their efficacy for conservation of natural resources and ensure environmentally sound development in the entire Indian Himalayan Region (IHR). As per the mandate, the Institute is engaged in maintaining the balance of intricate linkages between sociocultural, ecological, and physical systems that could lead to sustainability in the IHR. Apart from undertaking research and technology for development and demonstration on its own, the Institute has established linkages with some premier national- and international-level organizations committed to environment and development issues linked with the mountain regions. Toward addressing these objectives, this book is an outcome of the deliberations held during a workshop on "*Impacts of Global Change on the Dynamics of Snow, Glaciers and Runoff over the Himalayan Mountains, with particular reference to Uttarakhand*," organized by the Institute during 27–28 February, 2012. I am delighted to introduce the proceedings of the workshop in the form of a book entitled "*Dynamics of Climate Change and Water Resources of Northwestern Himalaya*," by GBPIHED and jointly published by the Society of Earth Scientists and Springer International Publishing AG, Switzerland. This publication attempts to understand the dynamics of climate change and water resources of Northwestern Himalaya. The Himalayan region, an integral part of the global ecosystem with its very rich repository of natural resources, serves as the water reservoir and a regulator of climate for the South Asian region. Climate change is a major concern in the Himalaya due to its potential impacts on the economy, ecology, and environment of the Himalaya as well as all areas downstream. The Himalayan region, most sensitive to global warming, shows stark impact of climate change on water resources and therefore needs dedicated efforts for continuous monitoring and advanced studies. I congratulate my colleagues and editors of this book, Dr. Rajesh Joshi, Scientist-D; Er. Kireet Kumar,

Scientist-G; and Dr. Lok Man S. Palni, Former Director of GBPIHED for putting their sincere efforts in bringing out this vital publication on very important interwoven issues of climate change and water resources. It is my sincere hope that this publication will inspire critical research on various innovative aspects of climate change and water resources of the Himalayan region.

Dr. P.P. Dhyani
Director
G.B. Pant Institute of Himalayan Environment and Development

Series Editor Foreword

Impending climate change has raised several questions and, in turn, initiated extensive research owing to its impact on society. The Himalayas, the northern rampart of the Indian subcontinent, fascinates geoscientists as it is the storehouse of freshwater resources for the future generation. Therefore, impact of climate change on the Himalayan region draws considerable attention. However, systematic data acquisition on the Himalayan region is a challenging task owing to topographic constraints and, therefore, whatever data are available on the topic must reach a wider spectrum of scientists for further research and analyses. Constant research is required to have an assessment on the socioeconomic impact of climate change. Global and regional cooperation is required to understand the actual impact of climate change on Himalayan glaciers by regular monitoring, and collation and interpretation of datasets. Therefore, the outcome of deliberations on 'Impacts of global change on the dynamics of snow, glaciers and runoff over the Himalayan Mountains' organized by GBPIHED, Almora, is compiled in this book.

The book is divided into three parts—Dynamics of Snow and Glaciers in North-West Himalaya; Assessment of Climate Change Patterns and Consequences of Changes and Flow Regime in order to understand the behavior of climate change in northwestern Himalaya. GLOFs is a new emerging area of study in glaciated terrains due to its possible disastrous impacts. The outcomes of melting glaciers are proglacial lakes and their increasing size and potential for outburst needs systematic study, particularly where there is a possibility for impact on life and property. The Kedarnath disaster is one such example. Further, the changing trend in the hydrological cycle on a regional or local scale is another area of research that invites the attention of geoscientists. Although scanty data is available, a concerted effort is still required. I am personally thankful to the editors of this volume for bringing out the available data on this important topic of societal issue. I am sure that the outcome of this book will enhance our efforts in the future by mutual collaborations and systematic data generation.

Satish C. Tripathi

Contents

Introduction 1
Rajesh Joshi, Kireet Kumar and Lok Man S. Palni

Part I Dynamics of Snow and Glaciers in North-West Himalaya

Variations in the Seasonal Snow Cover Area (SCA) for Upper Bhagirathi Basin, India 9
Rajesh Joshi, Kireet Kumar, Jibotosh Pandit and Lok Man S. Palni

Investigating Impacts of Global Change on the Dynamics of Snow, Glaciers and Run-off over the Himalayan Mountains 23
Kedar L. Shrestha

Identification of Glacial Lake and the Potentially Dangerous Glacial Lake in the Himalayan Basin 35
Sanjay K. Jain, Anil K. Lohani and Raj D. Singh

Assessment and Simulation of Glacial Lake Outburst Floods for Dhauliganga Basin in Northwestern Himalayan Region 45
Anil K. Lohani, Sanjay K. Jain and Raj D. Singh

A Model Study of Dokriani Glacier, Garhwal Himalaya, India 57
Argha Banerjee and R. Shankar

Part II Assessment of Climate Change Patterns

Critical Evaluation and Assessment of Average Annual Precipitation in The Indus, The Ganges and The Brahmaputra Basins, Northern India ... 67
Abul A. Khan, Naresh C. Pant, Anuj Goswami, Ravish Lal and Rajesh Joshi

Climate Change in the Northwestern Himalayas ... 85
Mahendra R. Bhutiyani

Aerosols and Temperature Rise in the Northwestern Himalaya, India ... 97
Jagdish Chandra Kuniyal

Measurement of Atmospheric Carbon Dioxide Levels at Dokriani Bamak, Garhwal Himalaya, India ... 115
Ashish Anthwal, Varun Joshi, Suresh C. Joshi and Kireet Kumar

Part III Consequences of Changes and Flow Regime

Hydrological Management of Glacial and Non-glacial River Systems ... 129
Kireet Kumar, S. Joshi, V. Adhikari, H. Sharma and T. Pande

Variable Response of Glaciers to Climate Change in Uttarakhand Himalaya, India ... 141
Dwarika P. Dobhal and Bhanu Pratap

Declining Changes in Spring Hydrology of Non-glacial River Basins in Himalaya: A Case Study of Dabka Catchment ... 151
Charu C. Pant and Pradeep K. Rawat

Chemical Characterization of Meltwater from East Rathong Glacier Vis-à-Vis Western Himalayan Glaciers ... 181
Brij M. Sharma, Shresth Tayal, Parthasarathi Chakraborty and Girija K. Bharat

Socio-economic Dimension of Snow and Glacier Melt in the Nepal Himalayas ... 191
Narayan P. Chaulagain

Author Index ... 201

Subject Index ... 203

About the Editors

Dr. Rajesh Joshi is presently working as Scientist-D at G.B. Pant Institute of Himalayan Environment and Development (GBPIHED), Almora. Dr. Joshi holds a Ph.D. degree in Applied Mathematics from G.B. Pant University of Agriculture and Technology, Pantnagar, India and has expertise in mathematical modeling using soft computing techniques. Dr. Joshi has over 10 publications to his credit, which include research papers, book chapters, peer reviewed reports, etc. He has experience in teaching (2 years) and research (7 years). His main research areas include soft computing techniques, environmental and hydrological modeling, climate impact studies, etc. Dr. Joshi has been a recipient of merit scholarship during his Ph.D. and the Young Scientist Award conferred by UCOST (DST), Dehradun. Also, he is a member of Society of Earth Scientists, India; International Association of Hydrological Science (IAHS), and Indian Association of Hydrologists (IAH).

Er. Kireet Kumar working as Scientist-G and Head WPM and KCB group of GBPIHED, Almora, has completed his M. Tech in Environmental Engineering from IIT, Kanpur, and has 15 years of experience in the field of glaciology. His main research interests include water resource management, glacier studies, soil and water conservation, and water quality management. He has over 40 publications including research papers, reviews, book chapters and has edited/co-authored books. He is involved in various projects leading to technical and policy documents on catchment area management, augmentation and management of spring water through spring sanctuaries, village environment action plan (VEAP), glacier studies and global positioning system surveys of landscape mapping, etc. He is also a Lead Fellow.

Dr. Lok Man S. Palni Former Director of GBPIHED, Almora, served at the Institute for nearly 20 years until his superannuation in May, 2013. Dr. Palni obtained his Ph.D. from the University of Wales, UK. He has nearly 42 years of experience in the broad area of Plant Sciences, and his other interests include environmental issues of the Himalayan region, rural development, and science and society, especially children centric programs. He has published over 300 research and review articles in peer-reviewed journals of international and national repute, book chapters as well as edited/authored books. Dr. Palni has been a recipient of merit scholarships and prestigious awards, e.g., Uttaranchal Ratna, Biodiversity Award (NASI), Shri Ranjan Memorial Award and is Fellow of the National Academy of Sciences, India (NASI), Allahabad (1997), National Academy of Agricultural Sciences (2011), the Indian National Science Academy (2012), and the International Society of Plant Morphologists, etc. He is a member of several high-powered scientific advisory boards, steering committees, and task force of various national and state level organizations. Presently he is with the Graphic Era University, Dehradun as Professor and Dean, and is also associated with the assessment work of Intergovernmental Science-PolicyPlatform on Biodiversity and Ecosystem Services (IPBES).

Introduction

Rajesh Joshi, Kireet Kumar and Lok Man S. Palni

Abstract The Himalayan region harbors plenty of water resources exploited by the populace of mountainous and downstream areas for domestic uses and other purposes. This region, source of supply to almost 80 % of the water resources and major rivers of North India, has profound influence on the climate and environmental front of this region. Indus and Ganges are the two major rivers in Western Himalayan region which directly impact the lives of a large population living in northern part of India, and even beyond the national boundaries. The Himalayas contain over half the permanent snow and ice-fields outside the polar regions. Because of the potential impacts of climate change on ecology and environment, the Himalayan region is considered as one of the most sensitive regions to global warming as change in climate has marked effect on water resources. Considering the importance of this "Water Tower" of South Asia, study of its water resources becomes imperative in context of changing climate. In the present book, attempts have been made to analyze the dynamics of climate change and water of the Northwestern Himalayan (NWH) region. In this publication various aspects related to dynamics of climate change and water resources including seasonal snow cover, glacier, melt runoff, rainfall, GLOFs, climate change, aerosols, atmospheric CO_2 level in glaciated catchment, hydrology of glacial and non-glacial river systems and springs, glacier retreat and mass balance, chemical characterization of glacier melt water, and socio-economic dimension of snow and glacier melt have been covered. The present book, an outcome of the deliberations held during

R. Joshi (✉) · K. Kumar
Watershed Processes and Management, G.B. Pant Institute of Himalayan Environment and Development, Almora, Uttarakhand, India
e-mail: dr.rajeshjoshi@gmail.com

K. Kumar
e-mail: kireetkumar@yahoo.com

L.M.S. Palni
Graphic Era University, Dehradun, Uttarakhand, India
e-mail: lmspalni@rediffmail.com

R. Joshi et al. (eds.), *Dynamics of Climate Change and Water Resources of Northwestern Himalaya*, Society of Earth Scientists Series,
DOI 10.1007/978-3-319-13743-8_1

the workshop organized by GBPIHED, also attempts to understand and estimate impacts of climate change on the dynamics of snow, glaciers, and runoff over the Himalayan Mountains and their consequences, both for the upland and downstream regions. The contents of the book have been summarized in the three sections (i) Dynamics of Snow in North–West Himalaya, (ii) Assessment of Climate Change Patterns, and (iii) Consequences of Changes and Flow Regime.

Keywords Climate change · Water resources · Snow cover · Glacier · GLOF · Aerosol · Hydrology · Springs · Northwestern Himalaya

Himalayan ranges, the source of fresh water supply and a perennial store house of ice, snow and permafrost as well as a vast repository of rich biodiversity, have always evoked profound interest in the global scientific community. In recent times, climate change has fuelled major research agenda to understand the processes and interactions operative in the region in order to approximate possible impacts. The Himalayan ranges house numerous glaciers and hundreds of small and big lakes, many of which are considered sacred. As per the inventory, there are 9,575 glaciers in the Indian part of Himalaya, out of which the Indus basin houses 7,997 and the Ganga Basin (including the Brahmaputra basin) has 1,578 glaciers located across five states of India, namely Jammu and Kashmir, Himachal Pradesh, Uttarakhand, Sikkim and Arunachal Pradesh. Himalayan rivers yield almost double the amount of water as compared to peninsular river. This is because glaciers and snow contribute important components of flow in these rivers (i) in years of deprived monsoon, (ii) during the lean summer and post-monsoon months; both of these factors help to reduce inter annual and inter seasonal variability, sustain water availability, hydropower generation, and agricultural production. Change trends in temperature and snow precipitation are bound to impact the hydrological cycle causing altered volume and timing of river runoff. At the global level there seems to be no homogeneous trend relating stream flows to temperature or precipitation changes. In fact it is also felt that water resource issues have not been adequately investigated in the climate change analyses and climate policy formulations.

The effects of climate change on stream flow and related variables are crucial for the planning of water resources and their management in time and space. Snow and glacier melt plays a crucial role, both in upstream as well as downstream areas. The downstream effects of changing water flow regimes in the large Himalayan rivers are largely, unknown. It is likely that these changes will have major impacts on downstream societies. IPCC in its fourth assessment report (AR4) has estimated that warming of the earth-atmosphere system is likely to change temperature and precipitation which may affect both the quantity and the quality of freshwater resources and various sectors (such as tourism, agriculture, forests, human health, industry, etc.) across the world. Furthermore, rising population and ever increasing pace of economic development will additionally enhance demands for fresh water in the changing scenario of reduced availability of resource. Therefore, quantitative estimates of the effects of climate on hydrology are essential for understanding, planning and management of water resource systems in the future.

In the above context, a 2 days national workshop on "Impacts of global change on the dynamics of snow, glaciers and runoff over the Himalayan Mountains, with particular reference to Uttarakhand" was organized by GBPIHED, Almora during 27–28 February, 2012. The aim of the workshop was to bring together leading experts working in the subject area to deliberate on issues related to alterations in the dynamics of Himalayan snow, glaciers and runoff vis-a-vis climate change so as to provide much needed science based information for identifying and implementing adaptation and mitigation strategies for sustainable development of the region. The present book, an outcome of the deliberations held during the workshop, attempts to understand and estimate impacts of climate change on the dynamics of snow, glaciers and runoff over the Himalayan mountains and their consequences, both for the upland and downstream regions. The contents of the book have been summarized in the following three parts.

1 Part I: Dynamics of Snow and in North–West Himalaya

The first chapter in this part presents an assessment of the variations in snow cover over space and time in the upper part of Bhagirathi river basin, Uttarakhand Himalaya. MODIS data have been used which provide repetitive coverage, and thus enable monitoring of snow variations at small time intervals. The use of MODIS data is one of the first attempts at snow cover monitoring in the study basin. This chapter narrates the importance of the work undertaken in monitoring of snow cover variations in the Bhagirathi river basin, and also highlights the details of MODIS satellite and the frequency of data that have been used in the work presented. The authors have also digitized the contours, at 40 m interval, in a GIS environment for preparing the digital elevation model and the altitudinal snow cover variations. They have attempted to assess the snow cover variations in different altitudinal zones across seasons, and have discussed seasonal and decadal variations.

The second chapter presents an approach for melt-runoff modeling which is quite meaningful for the prediction of meteorological and run-off parameters. Author has used different existing models using the data already generated, to predict the future meteorological and discharge scenarios, particularly in the Koshi basin (Nepal) and partly in the Hunza basin (Pakistan). The author has derived interesting future scenarios for Koshi basin that can lead to meaningful outcomes for studies in glacierized areas. The adopted approach and the methodology open new avenues of research and future activities in the high altitude glacierized basins of the Himalaya. The third chapter in this part, entitled "Identification of Glacial lake and the potentially dangerous glacial lake in the Himalaya basin", discusses the generic methodology used to identify the Glacial lakes. It is an important aspect under the climate change conditions that may give rise to the formation of large number of glacial lakes and GLOFs causing huge destruction. The authors have made mention of the identification of glacial lakes in the Himalayan basins. The next chapter on "Assessment and simulation of glacial lake outburst floods for a basin in Himalayan region" deals with the methodology of assessment of GLOF. A case study has been taken up to illustrate the methodology employed by using the MIKE 11 model. It has been argued that the mathematical models can indeed be very useful for this purpose; however, these require accurate longitudinal and cross-sectional data on the drainage systems using advanced technologies such as LIDAR scanning. The last chapter on "A model study of Dokriani glacier, Garhwal Himalaya, India" deals with one dimensional flow line model to simulate the glaciation process as well as the future behaviour of the glaciers. The model has been used on the Dokriani glacier and found to be satisfactory for predicting the observed changes in the glacier.

2 Part II: Assessment of Climate Change Patterns

Chapters under this theme cover different aspects of climate change patterns in the Northwestern Himalaya. The sparse data coverage and very few stations with significantly long term climatological observations are obvious limiting factors for conclusive research to arrive at a convincing picture of changing climate. However, authors tried to highlight diverse aspects of climate change issues using the data available at selected stations to draw conclusions about temporal change of local climate in the Himalaya. Low density of rain gauge stations especially in mountainous area, extreme variation in altitudes and large size of these basins forces adaptation of remote sensed data for estimation of average annual precipitation. Further, it has also been highlighted that recent augmentation of ground observations, supplemented by remotely sensed data can bring out the changes in the last few decades, with better spatial resolution.

First chapter in this part presents varying nature of average annual precipitation in three major river basins of India including the Indus, the Ganges and the Brahmaputra which constitute more than 50 % of the river discharge of India. Authors have used 11 years (2000–2010) Tropical Rain Measurement Mission (TRMM) generated radar precipitation raw data for estimation of the annual precipitation for the Indus, the Ganges and the Brahmaputra basins. The contoured distribution of precipitation indicates the orographic control as the primary factor on the summer monsoon precipitation in the Ganges and the Brahmaputra basins.

Next three chapters in this part deal with analyses and/or observations of climatologically important parameters, such as temperature, rainfall, CO_2 levels, and aerosols in the atmosphere and their optical properties. Contents of these chapters will be useful for the readers as they summarize some of the climate change related conclusions in the Himalayan region. The chapter on "Climate Change in the northwestern Himalaya" is a comprehensive review of the science of climate change in the northwest Himalaya. The author has used long term

(1866–2006) climatological data of three stations in the region and provided details of the recent data network with reasonably good spatial distribution. The chapter further describes the general climatology of the region. The next chapter on Aerosols presents details of the measurements at an important location in Himachal Pradesh in Northwestern Himalaya. The last chapter that describes the atmospheric CO_2 levels at Dokriani Bamak glacier is both unique and interesting, and deals with a comparative estimate of annual mean CO_2 concentration in Dokriani Bamak of Garhwal Himalaya, along with the global average values. The patterns in respect of diurnal variations of CO_2 levels across different months are reflective of combined effects of biogenic and meteorological factors. The authors have tried to explain the possible cause of high CO_2 levels in the relatively cleaner areas, such as glaciers of the Himalayan region.

3 Part III: Consequences of Changes and Flow Regime

The first chapter in this part investigates the specific features of hydrological behaviour of glacial and non-glacial river systems. For a case study, authors have taken one glacial basin (the Gangotri glacier-one of the largest glaciers in Garhwal region of Indian Himalaya) and a non-glacial basin (the upper Kosi basin in Kumaun region of Indian Himalaya) of Uttarakhand state in India. Glacial basins are characterized as high energy landforms with less biotic activities, whereas the non-glacial basins have gentle slopes and are subjected to more intense biotic activities. Hydrological responses of the studied basins confirm the role of these characteristics.

The glaciers are fragile and dynamic in nature and influence the climate system (e.g. albedo feedback) and hence are key indicator of climate change. The reduction in mass, volume, area and length of glaciers are considered as clear signals of a warmer climate. The second chapter deals with variable response of glaciers to climate change in Uttarakhand Himalaya, India. In this chapter, the authors have presented the results of a detailed mapping campaign and ground-based measurements for terminus retreat, area vacated and mass/volume change carried out on few glaciers for the period 1962–2010. The study shows continuous negative mass balance on Tipra, Dunagiri, Dokriani and Chorabari glaciers during last three decades. The study shows that the glaciers of Uttarakhand Himalaya are under substantial thinning (mass loss) and reduction of length and area in the present climate conditions.

The third chapter on "Declining changes in spring hydrology of non-glacial river basins in Himalaya" deals with a very important aspect of natural springs in the Himalayan areas, which are a major source of water for the local communities. This chapter, through a case study of Dabka catchment in Kumaun region of Uttarakhand, has tried to establish that springs which exist along the thrust/fault planes and fluvial deposit areas are perennial, and most of those that exist along fracture/joints and shear zones are non-perennial. The authors have tried to establish a relationship on springs' yield and geology based of the case study presented.

The next chapter of the part presents an analysis of chemical characterization of glacier melt water vis-à-vis western Himalayan meltwater streams. As a case study, snow samples of late winter season from Rathong Glacier and its pro-glacial stream Rathong Chhu were analyzed to study the chemical composition, weathering, and geochemical processes in ice and meltwater at high altitudes. Analyses shows that enrichment of samples with NO_3^- and NH_4^+ suggests scavenging of HNO_3 present in the atmosphere is a major contributor for these ions. Lastly the chapter six of this part deals with the socio-economic dimension of snow and glacier melt in the Nepal Himalaya. The author has investigated possible impacts of change in the runoff due to changed climate, and has discussed the resulting consequences in the downstream areas of Koshi river basin in Nepal.

The melting of snow and glaciers and subsequent changes in water regime have multi-facet impacts on society and economy because of direct linkage of water with people, ecosystem, economy and society. It is therefore, concluded that the impacts of climate change on runoff regime and other water resources would widen the gaps between water supply and demand disproportionately on marginalized and subsistent communities and the economic units which are directly dependent on natural system.

We wish to thank the authors for contributing research articles for this book, and for active participation and deliberations in various technical sessions in the workshop. Uttarakhand State Council for Science and Technology (UCOST), Dehradun and Asia Pacific Network for Global Change Research (APN), Japan are thanked for assistance and being co-sponsors of the workshop. Thanks are also due to Dr. S.C. Tripathi, Series Editor, SES-Springer series for his support in publication of this book. The editors wish to thank the reviewers: Prof. G.B. Pant, former Director, IITM, Pune; Prof. A.K. Gosain, IIT, Delhi; Dr. Sharad Jain, NIH, Roorkee, and Dr. A.K. Tangri, UP-RSAC, Lucknow who have critically gone through the papers and provided critical suggestions for review of the papers. Their suggestions and comments greatly helped the authors to improve their respective contributions. Finally Director, G.B. Pant Institute of Himalayan Environment and Development, Kosi-Katarmal, Almora, is thanked for providing necessary support for the work.

Part I
Dynamics of Snow and Glaciers in North-West Himalaya

Variations in the Seasonal Snow Cover Area (SCA) for Upper Bhagirathi Basin, India

Rajesh Joshi, Kireet Kumar, Jibotosh Pandit and Lok Man S. Palni

Abstract Satellite based remote sensing is a convenient tool for the study of cryosphere that allows to carry out investigations over large and inaccessible areas. The present investigation has been carried out to monitor seasonal variation in the Snow Cover Area (SCA) for the upper Bhangirathi basin, located in the Garhwal region of Indian Himalaya. This analysis has been done using Moderate-Resolution Imaging Spectroradiometer (MODIS) satellite data for the past 11 years (2000–2010); the temporal snow cover being derived using the Normalized Difference Snow Index (NDSI). The entire study basin has been divided into nine elevation zones, on the basis of Digital Elevation Model (DEM), for estimating the SCA for each zone. Zones 1–9 cover different elevation ranges: (1) above 6,500 m, (2) between 6,000 and 6500 m, (3) 5,500–6000 m, (4) 5,000–5500 m, (5) 4,500–5000 m, (6) 4,000–4500 m, (7) 3,500–4000 m, (8) 3,000–3500 m, and (9) below 3,000 m. Mann Kendall and linear regression methods have been employed to identify trends in the SCA during the period 2000–2010. The snow cover depletion analysis depicts a shift in the duration of ablation and accumulation during the study period in the basin. The analysis indicated 13–21 % increase in SCA in the middle elevation zones (4 and 5) and 2–9 % decline in SCA in the lower elevation zones during autumn. SCA was found to increase across all the elevation zones in winter; the rate of increase was particularly high (14–21 %) in the lower elevation zones as compared to higher (2–3 %) and middle elevation zones (4–10 %). Similarly, an increase of 2–3 % in the higher elevation zones, 6–14 % increase in the middle elevation zones and 2–6 % decline in the lower elevation zones was observed in respect of SCA during spring. However, no significant variation in SCA was observed during the summer season. Decadal variation in SCA showed mean annual

R. Joshi (✉) · K. Kumar · J. Pandit
Watershed Processes and Management, G.B. Pant Institute of Himalayan Environment & Development, Almora, Uttarakhand, India
e-mail: dr.rajeshjoshi@gmail.com

L.M.S. Palni
Graphic Era University, Dehradun, Uttarakhand, India

R. Joshi et al. (eds.), *Dynamics of Climate Change and Water Resources of Northwestern Himalaya*, Society of Earth Scientists Series,
DOI 10.1007/978-3-319-13743-8_2

increase of 8–15 % in the middle elevation zones (3–5). In the lower elevation zones (<4,500 m), mean annual SCA showed increase of 11–14 % between 2000 and 2005, followed by 6–8 % decrease in the upper Bhagirathi Basin.

Keywords Seasonal snow cover · SCA · MODIS · NDSI · Upper Bhagirathi Basin · Garhwal Himalaya

1 Introduction

Technological advancements in the space science have provided steadily increasing satellite platforms that can be used to study complex physical processes in the earth-atmosphere system. In particular for glaciological studies, the study areas are often inaccessible with harsh climatic conditions; satellite remote sensing has proven to be of special value for the year round real-time observations. The large spatial coverage of remotely sensed data enables real-time monitoring and process studies over large areas and help to understand various processes on a regional, continental, or even at a global scale. Such products are particularly important as they assist in the interpretation and analysis of patterns of global change. Furthermore, on a smaller scale, satellite remote sensing can be crucial, both for basic understanding of processes (e.g., glacier velocities from interferometry products), and snow cover and glacier retreat monitoring (Hall and Martinec 1985; Foster et al. 1987 and Massom 1991). Especially for the data-deficit regions, such as Himalaya, satellite derived SCA information is perhaps the best routinely available input for snowmelt runoff estimates (Rango 1985).

In the recent past, satellite based remote sensing techniques have been widely used globally for investigating glacier fluctuations and snow cover changes (Kulkarni and Rathore 2003). Using these techniques, Menon et al. (2010) have estimated the annual snow cover of the Hindu Kush-Himalayan (HKH) region that showed a decline of 16 % during 1990–2001, the contribution of enhanced black carbon to the reported decline in SCA being close to 36 %. Changes in the snow cover in HKH region are primarily on account of inter-annual variations in the circulation pattern (Gurung et al. 2011). In addition, topographic differences play an important role in climatic variations and spatial variability of snow cover (Immerzeel et al. 2009). The western Himalayan region accounts for a higher average snow cover due to higher mean elevation and the influence of winter westerlies (Bookhangen and Burbank 2010). Similarly, studies conducted in the western Himalaya (Kriplani et al. 2003) reveal that the timing of snow cover peak differs due to the influence of different weather systems. For example, in the Kashmir valley, the peak snow cover is generally observed in February (Negi et al. 2009) whereas in the Bapsa basin of Himachal Pradesh the same is observed during March end. The variation in seasonal snow cover significantly affects the snow line as well. The recent study by Kulkarni (2010) in Chhota Shigri glacier clearly indicates a change in snow line from 4,900 to 5,200 m (from late 1970s to present). These changes may be attributed to large depletion of snow cover in early winter, i.e., from October to December. Various satellite remote sensing based methods exists for measuring snow and glacial ice properties. These methods have steadily improved with the use of new satellite sensors, like synthetic aperture radar (SAR) during the past few decades. This has led to the development of powerful methods, such as SAR interferometry for estimating glacier velocity, generation of DEM for ice sheets, and snow cover mapping. Common methods include satellite-derived determination of surface albedo, snow volume and extent, surface temperature, glacier velocities and extent, and ice sheet topography.

The Moderate-Resolution Imaging Spectroradiometer (MODIS) belongs to a new class of satellite sensors with greatly improved capabilities. This imaging spectrometer is able to record continuous spectral range, in contrast to multispectral scanners. Since February 2000, daily global snow cover maps are being generated from the data collected by the MODIS satellite. NASA's MODIS began collecting data from the Terra (formerly known as EOS-AM1) spacecraft on February 24, 2000 and from the Aqua (formerly known as EOS-PM1) spacecraft on June 24, 2002. MODIS views the earth's entire surface daily from high latitudes to every other day at low latitudes. It has 36 spectral bands in the visible, near-infrared, short-wave infrared and thermal portions of the electromagnetic spectrum (Justice et al. 1998). The Distributed Active Archive Center (DAAC) at the National Snow and Ice Data Centre (NSIDC) currently distributes the snow cover products produced from MODIS satellite (Scharfen et al. 1997; Hall et al. 2000). Their global extent and daily coverage enables the MODIS snow products to augment existing remote-sensed derived continental and regional scale snow cover maps and provide high resolution snow cover information for areas of the world where snow cover maps are currently not produced. In the present paper MODIS Terra snow products for 2000–2010 have been used to study seasonal variation in snow cover area in the upper Bhagirathi Basin.

2 Study Area

The study has been carried out for the upper Bhagirathi Basin, located in the Uttarkashi district of Garhwal region in Uttarakhand, India. The river Bhagirathi, originating from Gaumukh (snout of Gangotri glacier), joins with Alaknanda at Devprayag to form the mighty Ganges. The upper Bhagirathi valley starts at Gaumukh and ends at Gangotri after a distance of approximately 18 km. The valley is enchanting and has many scenic spots, many of its enchanting features include white-water streams, snow-clad high mountains, milky glaciers, conifer forests and alpine meadows. The total glacierized area of the valley is about 258.56 km^2, approximately 34 % of the total glacierized area of Bhagirathi catchment. The catchment area of Bhagirathi (upstream to Gangotri) falls in the mountainous Higher and Trans-Himalayan region. Total area of the upper Bhagirathi Basin is approximately 992 km^2. Location map of the study area is shown in Fig. 1. The basin has socio-economic importance as number of hydropower project have been established or, are, planned in the downstream areas where the stream flow influences the livelihood of people inhabiting the

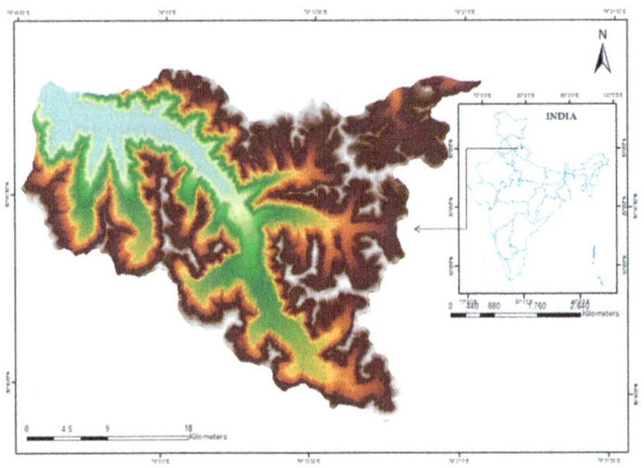

Fig. 1 Location map of the upper Bhagirathi Basin

region. The streamflow is mainly regulated by snow and glacier melt runoff at high altitude areas of the basin. Proper estimation of seasonal snow cover and SCA, on an annual basis, is important for a number of issues, such as water management.

3 Methodology

MODIS Terra 8-day maximum snow products for the past 11 year (2000–2010) have been used in this study were obtained from the DAAC, NSIDC of NASA. Temporal filtering method, as proposed by Hall et al. (2010), has been used for cloud masking in the data products. The MODIS snow cover products are generally available in Hierarchical Data Format of Earth Observation System (HDF-EOS). Application of ERDAS software is used to apply geometric correction to re-project the original sinusoidal file to geographic lat-long projection, where nearest neighbor re-sampling method is adopted. Model maker tool of the software is then used to classify a geometrically corrected MODIS data and to classify the DEM of the basin, and finally the interpreter tool is used to superimpose classified DEM and the MODIS snow product so as to obtain intersection for estimating SCA. In this way, Geographic Information System (GIS) tool is used to customize the automatic extraction of snow covered/free area under different elevation zones for the basin under study. Approximately 400 scenes were analyzed to generate long term time series for snow covered area. Daily values for snow cover were obtained from 8 day composite snow cover using interpolation technique.

3.1 Hypsometric Analysis of Upper Bhagirathi Basin

In the mountainous areas, hydrological and meteorological conditions are related to elevation; therefore, the study basin was divided into 9 elevation zones with an elevation difference of 500 m, using a 40 m resolution digital elevation model (DEM). Such a division of a catchment into elevation zones helps to give better estimates of zone-wise snow cover/free area that finally helps in determining the melt runoff contribution for each zone. The DEM of the study area was used for the preparation of elevation map (Fig. 2) and area-elevation curve using Geographical Information System (GIS).

The GIS technique was utilized to customize the tool for automatic extraction of area under different elevation zones for the upper Bhagirathi Basin. Using this tool, area-altitude distribution and the summary table has been generated for further analyses to generate hypsometric curve (Fig. 3) for the study basin. The distribution of basin area along with altitudinal variation indicated that the upper Bhagirathi Basin lies within the elevation range 2,600–7,200 m (zones 1–9), and that more than 70 % of the basin area lies between zones 3 and 5.

Further, it also showed that 86 % area falls within middle (zones 3–5) and lower (zones 6–9) elevation zones, where as only 14 % in the higher elevation zones (1 & 2). The summary of basin area under each elevation zone has been presented in Table 1.

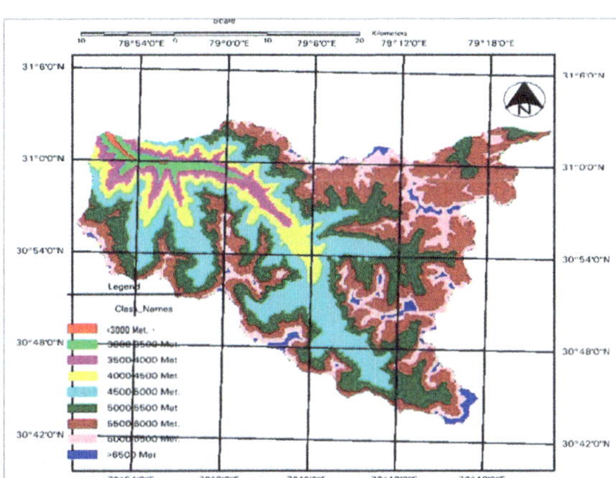

Fig. 2 Elevation map of the upper Bhagirathi Basin, delineated at 500 m intervals

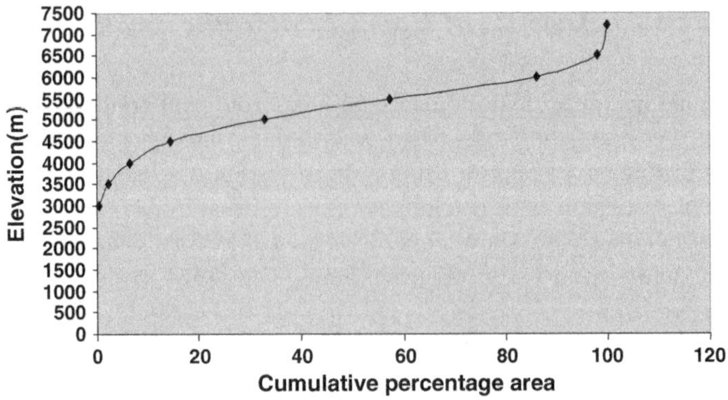

Fig. 3 Area altitude distribution curve for the upper Bhagirathi Basin

Table 1 Area-elevation characteristics of upper Bhagirathi Basin

Elevation zone	Elevation range (in m)	Area (in km^2)	Area as % of total upper Bhagirathi Basin	Cumulative area (%)	Categorization of zones
Zone 9	<3,000	3.60	0.36	0.36	Lower zone
Zone 8	3,000–3,500	19.63	1.97	2.34	
Zone 7	3,500–4,000	42.22	4.25	6.59	
Zone 6	4,000–4,500	79.05	7.96	14.56	
Zone 5	4,500–5,000	181.03	18.25	32.82	Middle zone
Zone 4	5,000–5,500	243.26	24.52	57.34	
Zone 3	5,500–6,000	287.12	28.94	86.29	
Zone 2	6,000–6,500	117.70	11.86	98.15	Higher zone
Zone 1	6,500>	18.25	1.84	100	
	Total	991.89			

3.2 MODIS Snow Cover

MODIS Terra 8 day composite snow cover products (MOD10A2) have been used to derive snow cover information for the basin. Each MOD10A2 file consists of single tiles of 500 m gridded geo-referenced cells covering 1,200 × 1,200 km area. From MOD10A2 products, the presence or absence of snow, and cloud cover in each pixel was estimated using the SNOWMAP, the MODIS snow mapping algorithm. In this algorithm, surface reflectance and the cloud mask are utilized as basic inputs to MODIS swath products in order to construct the daily and 8 day composite snow cover products (Hall et al. 1995). The normalized difference snow index (NDSI) method was used for the identification of a snow pixel. This method was useful for the identification of snow and ice, and for separating snow/ice and most of the cumulus clouds. The NDSI served as a measure of the relative magnitude of the characteristic reflectance difference between the visible and short-wave infrared (SWIR) reflectance of snow. The NDSI, employing MODIS bands 4 (0.545–0.565 μm) and band 6 (1.628–1.652 μm), was used as the primary snow-classification criteria:

$$NDSI = \frac{(Reflectance\ in\ Band\ 4 - Reflectance\ in\ Band\ 6)}{(Reflectance\ in\ Band\ 4 + Reflectance\ in\ Band\ 6)}$$

MODIS pixels with NDSI value greater than or equal to 0.40 are considered to represent snow. In general, snow is characterized by higher NDSI values in comparison to other surface types.

3.3 Historical Trend Analysis

Non-parametric Mann-Kendall and linear regression methods have been used to identify the trends in seasonal snow cover. Since Mann-Kendall method did not reflect any statistically significant trend in the seasonal snow cover, linear regression method was adopted to identify the trend in SCA across seasons during 2000–2010; it showed that the mean annual SCA was increasing across all elevation zones during the study period. The deduced trends have been discussed in Sect. 4.3.

4 Results and Discussions

Seasonal snow cover, a vital reservoir of water in many parts of the world, is an important hydrological phenomenon, which significantly contributes to river discharge as well as affects biotic components and water quality in river basins. To understand seasonal variation of SCA in the upper Bhagirathi Basin, data were analyzed for four different seasons, namely autumn (October–November), winter (December–March), spring (April–June), and summer (July–September).

4.1 Snow Cover Depletion Analysis

The snow depletion curve depicts accumulation and ablation behavior of seasonal snow and serves as an important input to the hydrological models for estimating regional runoff of the snow and glacier melt. The SCA for individual elevation zones (1–9) were plotted against elapsed time to construct depletion curves. Snow depletion curves plotted for the ablation period (i.e., May–September) have been shown in Fig. 4, which clearly depict a shift in the duration of ablation

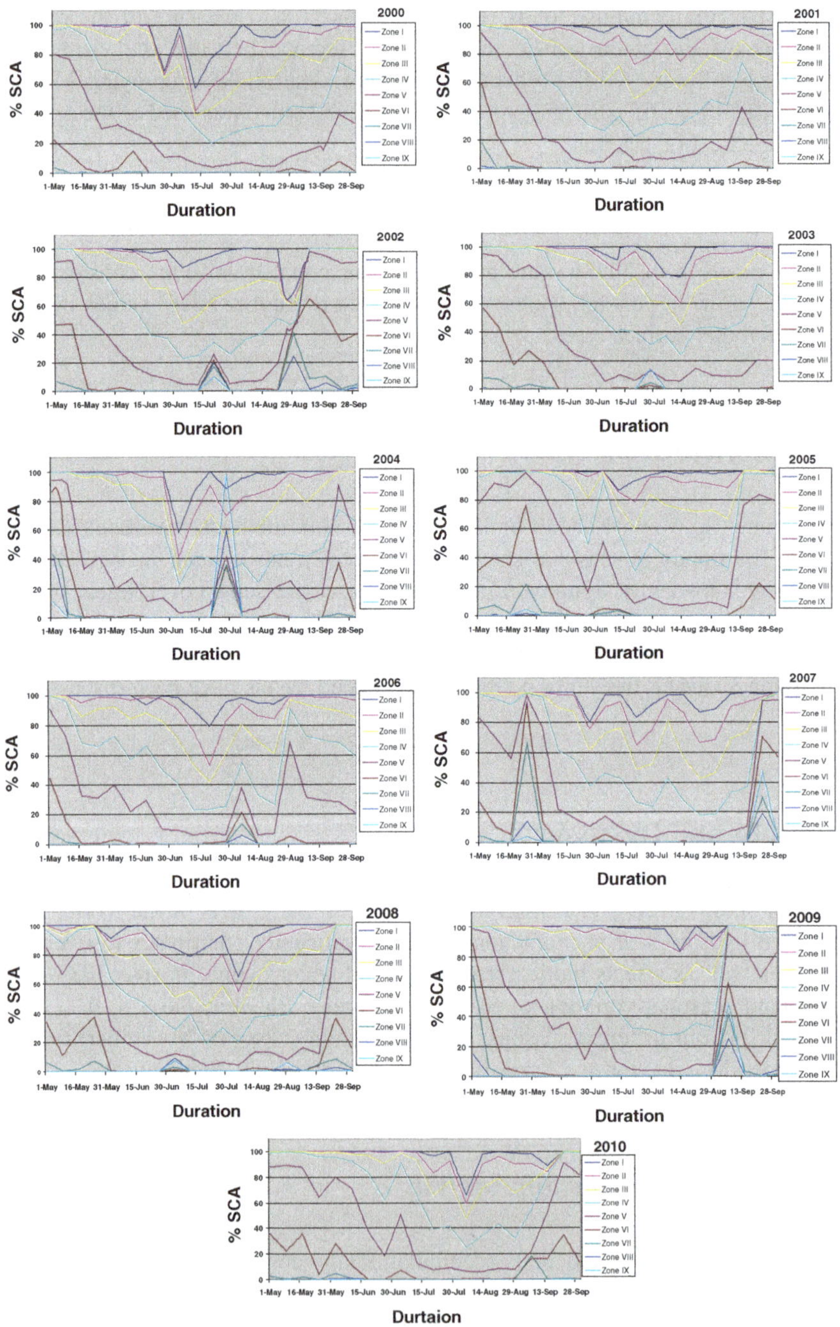

Fig. 4 Snow cover depletion pattern in different elevation zones during 2000–2010

and accumulation over study period of 11 years. Further, it was observed that the ablation of snow in the lower and middle elevation zones, which in earlier times used to begin by mid of May, has started to occur in the first week of May. Similarly, the accumulation of snow, that used to start by first week of September, is now seen to take place during the end of September. This indicates a clear shift in the ablation period towards early spring by around 2 week, whereas the accumulation of snow is seen to be delayed by almost 2–3 weeks in the lower and middle elevation zones. Consequently the total duration of ablation period appears to have extended by almost a month over the study period. The observed extension of ablation period causes greater exposure of snow and glacier covered areas that results in, both a shift towards earlier spring melt as well as higher snow and ice melt.

In line with the observations in HKH region (Gurung et al. 2011), the plotted curves show that the duration of ablation in higher zones has also increased, and the beginning of snow depletion is delayed with the elevation across the basin. At middle and lower latitudes, ablation happens through melting caused by the direct heat supplied by the solar radiation. At higher latitudes, the summer temperature may not rise above freezing; under such circumstances, ablation occurs mainly through sublimation when the snow converts to water vapor directly, without a liquid phase. Accumulation and ablation play a pivotal role on impacting climate-influenced dynamics on snow and ice that ultimately determine the overall water balance and seasonal fluxes in a given glacier's budget. These changes, therefore, may be attributed to the changing climate and inter annual variations in the circulation pattern (Gurung et al. 2011). Since, the timing and volume of spring runoff is affected by many variables, including the accumulated snow pack depth, its distribution and the retention of snow pack over time. Therefore, assessment of snow accumulation and depletion is essential for the management of water resources in the region where snow forms a large part of the annual precipitation.

4.2 Changes in Seasonal SCA

Seasonal variation in snow covered area in different altitudinal zones has been studied for various seasons (i.e., autumn, winter, spring, and summer) during 2000–2010, and the observations presented in Fig. 5. During the study period, SCA in the autumn season showed varying levels of increase in the higher and middle elevation zones with particularly higher rate of increase (13–21 %) in the middle zones (4 and 5); in the lower zones (6–9), 2–9 % decline in SCA was observed. In winter, SCA showed an increase during the study period across all elevation zones. During winter, the rate of increase of SCA recorded for the higher zones was low (2–3 %) in comparison to the middle (4–10 %)

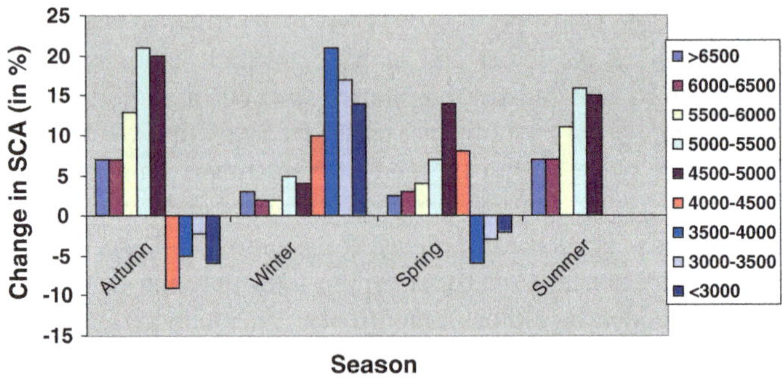

Fig. 5 Percentage change in SCA in different seasons during 2000–2010

and lower elevation zones (14–21 %). The analyses revealed that, in spring, 2–3 % and 6–14 % increase in the higher and middle zones, respectively, and 2–6 % decline in SCA in the lower elevation zones. In summer, non-significant SCA increase in the higher zones (approx. 7 %) and significantly greater increase (11–13 %) in the middle elevation zones was observed. Since, solid precipitation events rarely take place in the lower elevation zones during summer SCA at such altitudes was found to depict no or almost negligible variations.

4.3 Decadal Variation in SCA

The decadal variation in snow cover change in the upper Bhagirathi Basin, for 2000–2010, has been estimated using mean annual snow cover area derived from standard Terra 8-day MODIS snow products. The observed variations (as presented in Fig. 6a–d) showed an overall increase in the mean annual SCA across all elevation zones of the basin.

The observed trends in mean annual SCA was, however, statistically not significant; the mean annual values for three different time periods (i.e., for the years 2000, 2005 and 2010) were further analyzed to investigate temporal variation in SCA, and the results are plotted in Fig. 7a, b). The plots clearly show that over a span of 11 years, the SCA did change in the basin depending upon the elevational ranges in the basin. In the higher elevation zones, variation in SCA during 2000–2010 was very small (almost 3 %). However, in the case of middle elevation zones (3–5), the annual mean SCA increased by 8–15 %. In the lower elevation zones (6–9), a completely different pattern of SCA were observed; while an increase in SCA was seen from 2000 to 2005, and a steady depletion recorded thereafter. The analysis showed that in the lower elevation zones, the mean annual SCA first increased from the year 2000 to 2005 by 11–14 % and then showed a decline from 2005 to 2010 by 6–8 %.

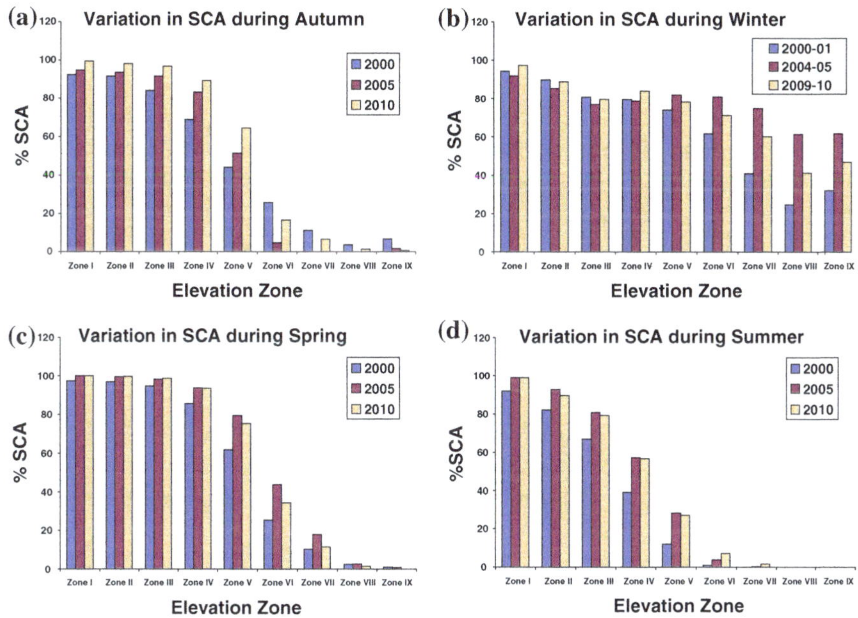

Fig. 6 Seasonal variation in SCA in **a** autumn, **b** winter, **c** spring, and **d** summer during 2000–2010

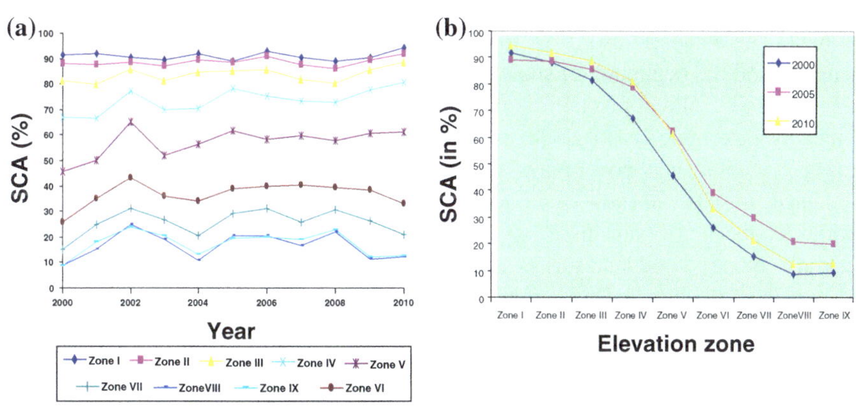

Fig. 7 **a, b** Decadal variation in SCA during 2000–2010

5 Conclusion

Snow is an important component of the cryosphere, and the study of snow trends is essential for understanding regional climate change and managing water resources. In this paper we have estimated variations in the seasonal snow cover for the upper

Bhagirahi Basin located, in Garhwal Himalaya of Uttarakhand, India. For this purpose, Terra 8-day snow cover products of MODIS satellite, for the past 11 years (2000–2010), were utilized. The temporal snow cover was derived using the NDSI method. With the help of a DEM, the entire study basin was divided into nine elevation zones, and the SCA for each zone was estimated. Mann Kendall method and linear regression tool were employed to identify trends in the SCA during the study period. The snow cover depletion analyses depict a shift in the duration of ablation and accumulation over the past 11 years. Based on the analyses, during 2000–2010, 13–21 % increase in SCA in the middle elevation zones (3–5) and 2–9 % decline in the lower zones was observed in autumn. In winter, SCA was found to increase across all elevation zones, with particularly higher rate of increase (14–21 %) in the lower elevation zones, as compared to the observed increase in the higher (2–3 %) and middle zones (4–10 %). Similarly, an increase of 2–3 % in the higher zones, 6–14 % in the middle zones and a decline of 2–6 % in the SCA of lower elevation zones SCA were observed in spring, whereas no significant variations were observed during summer SCA. Decadal variation in SCA showed that in the middle elevation zones (3–5), mean annual SCA showed increase by 8–15 %. In the lower elevation zones (6–9), the mean annual SCA showed an increase of 11–14 % between the years 2000 and 2005, followed by a decrease of 6–8 % in the upper Bhagirathi Basin.

Because of the shift in the ablation period towards early spring and delay in the accumulation of snow in the lower and middle elevation zones, the total duration of ablation period appears to have extended by almost a month over the study period. The extension of ablation period cause greater exposure of snow and glacier that consequently result in, both a shift towards earlier spring melt as well as higher snow and ice melt. The analyses also show that duration of ablation in higher zones has also increased and the beginning of snow depletion is delayed. Accumulation and ablation of snow and ice play a pivotal role on impacting climate-influenced dynamics which ultimately determine seasonal fluxes in glacier mass balance and the overall water budget. Seasonal snow cover plays an important role in the earth-atmosphere system, as changes in seasonal snow cover, ice and permafrost can also affect climate through various feedback mechanisms. Snow and ice surface have much higher albedo in comparison to other natural land surfaces, and are, therefore, capable of reflecting a greater percentage of the incident solar radiation. Hence, decline in snow cover may lead to an increase in the amount of solar radiation absorbed by the earth, therefore enhancing the global warming process. Seasonal snow cover being a vital water resource for irrigation, hydroelectric power generation, agriculture, and potable water, any change in the snow cover is likely to result in changed river flow regime, and would thus affect water availability in upstream and downstream areas of Himalayan region.

References

Bookhangen B, Burbank DW (2010) Towards a complete Himalayan hydrological budget: spatiotemporal distribution of snowmelt and rainfall and their impact on river discharge. J Geophy Res 115. doi:10.1029/2009JF001426

Foster JL, Hall DK, Chang ATC (1987) Remote sensing of snow. Eos Trans Am Geophy Union 68(32):682–684

Gurung DR, Kulkarni AV, Giriraj A, Aung KS, Shrestha B, Srinivasan J (2011) Changes in seasonal snow cover in Hindu Kush-Himalayan region. Cryosphere Discuss 5:755–777
Hall DK, Martinec J (1985) Remote sensing of ice and snow. Chapman and Hall, New York
Hall DK, Andrew BT, James LF, Alfred TCC, Milan A (2000) Inter-comparison of satellite derived snow-cover maps. Ann Glaciol 31:369–376
Hall DK, Riggs GA, Foster JL, Kumar SV (2010) Development and evaluation of a cloud-gap-filled MODIS daily snow-cover product. Remote Sens Environ 114:496–503
Hall DK, Riggs GA, Salomonson VV (1995) Development of methods for mapping global snow cover using moderate resolution imaging spectroradiometer data. Remote Sens Environ 54:127–140
Immerzeel WW, Doorgers P, de Jong SM, Bierkens MFP (2009) Large scale monitoring of snow cover and runoff simulation in Himalayan river basins using remote sensing. Remote Sens Environ 113:40–49. doi:10.1016/j.rse.2008.08.010
Justice CO, Vermote E, Townshend JRG, DeFries R, Roy DP, Hall DP, Salomonson VV, Privette JL, Riggs G, Strahler A, Lucht W, Myneni R, Knyazikhin Y, Running SW, Nemani RR, Wan Z, Huete A, van Leeuwen W, Wolfe RE, Giglio L, Muller JP, Lewis P, Barnsley MJ (1998) The moderate resolution imaging spectroradiometer (MODIS): land remote sensing for global change research. IEEE Trans Geosci Remote Sens 36:1228–1249
Kriplani RH, Kulkarni AV, Sabade SS (2003) Western Himalayan snow cover and Indian monsoon rainfall: a re-examination with INSAT and NCEP/NCAR data. Theor Appl Climatol 74(1–18). doi:10.1007/s00704-002-0699-z
Kulkarni AV (2010) Monitoring Himalayan cryosphere using remote sensing techniques. J Indian Inst Sci 90(4):457–469
Kulkarni AV, Rathore BP (2003) Snow cover monitoring in Bapsa Basin using IRS WIFS data. Mausam 54(1):335–340
Massom R (1991) Satellite remote sensing of polar regions. Lewis Publishers (CRC Press), Boca Raton
Menon S, Koch D, Beig G, Sahu S, Fasullo J, Orlikowski D (2010) Black carbon aerosols and the third polar ice cap. Atmos Chem Phys 10:4559–4571
Negi HS, Kulkarni AV, Semwal BS (2010) Estimation of snow cover distribution in Beas basin, Hindian Himalaya using satellite data and ground measurements. J Earth Syst Sci 118(5):525–538
Negi HS, Thakur NK, Kumar R, Kumar M (2009) Monitoring and evaluation of seasonal snow cover in Kashmir valley using remote sensing, GIS and ancillary data. J Earth Sys Sci 118:711–720
Rango A (1985) Assessment of remote sensing input to hydrologic models. Water Resour Bull 21:423–432
Scharfen GR, Hall DK, Riggs GA (1997) MODIS snow and ice products from the NSIDC DAAC. In: Proceedings of the SPIE, San Diego, CA, pp 143–147

Investigating Impacts of Global Change on the Dynamics of Snow, Glaciers and Run-off over the Himalayan Mountains

Kedar L. Shrestha

Abstract Himalayan rivers like Ganges and Indus are the source of fresh water and livelihood for millions living in the mountains and the downstream regions. The Himalayan glaciers and snow together with the base flow from the high mountains contribute important components of flows rendering the rivers perennial in nature. As water resources are inextricably linked with climate and other anthropogenic changes, global change is quite likely to alter significantly the flow patterns of the Himalayan rivers with consequent impacts on the economy in terms of water availability, food security and hydropower generation. Investigations on global change impact assessment on Himalayan hydrology involves the evaluation of changes in the snow and glacier dynamics as well as runoff resulting from changes in climatic, hydrological and local anthropogenic parameters. Such an ongoing investigation carried out in a few selected watersheds in the Himalayan region under a regional collaborative effort is briefly described and some results thus obtained are also presented and discussed.

Keywords Himalayan rivers · RCMs · Snow and glacial melt runoff

1 Introduction

Himalaya, the abode of snow with the third largest reservoir of snow and ice after Arctic/Greenland and Antarctic regions, is the source of all the major perennial rivers in Asia and provides fresh water to billions of people living in and around the region. Glaciers and snow in the Himalayas contribute important components of flows in the Himalayan rivers like Indus, Ganges and Brahmaputra in the years of poor monsoon and in reducing inter-annual and inter-seasonal variability during the lean summer and post monsoon months.

K.L. Shrestha (✉)
Institute for Development and Innovation, Lalitpur, Nepal
e-mail: klshrestha1@ntc.net.np

R. Joshi et al. (eds.), *Dynamics of Climate Change and Water Resources of Northwestern Himalaya*, Society of Earth Scientists Series,
DOI 10.1007/978-3-319-13743-8_3

Water being an essential resource for all forms of life on our planet, variations in hydrologic regimes often have serious consequences and, thus, are a potential threat for society. Contemporary global and regional climate and other environmental changes pose immense challenge to Himalayan water resources management due to high spatial and temporal variation of resource endowment, and upstream-downstream linkages as a result of high degree of interrelationship among water uses and users as well as their transboundary nature. It is therefore necessary to provide best possible information regarding future changes in the Himalayan hydrological cycle in such a way as to enable responsible decision makers to find possible strategies to mitigate or adapt to global change for the sustainable development in the region.

While global warming obviously is likely to impact the Himalayan snow and glaciers, the Fourth Assessment Report (AR4) of the Intergovernmental Panel on Climate Change (IPCC) released in 2007 (IPCC 2007) pointed out to the dearth of such impact studies and stated that *the high mountains of Asia till then remained a "white spot". Furthermore, based on the expected increase in global temperature in the new millennium, it also included a statement that all glaciers in the Himalayas could disappear by* 2035. While the second statement after considerable discourse was altered into a new statement (IPCC 2010), there appeared, subsequent to the release of the report, a surge in the studies and research activities on impacts of climate change on snow and glacier in the Himalayas as well as on their consequences on the runoff of the Himalayan perennial rivers.

Recent studies based on satellite imageries on the snow cover as well as the glacier extent and sizes in the Himalayan region clearly indicated a declining trend in all these parameters, although such trends were found to vary across the region (Gurung et al 2011; Bajracharya and Shrestha 2011). Global change is thus quite likely to alter significantly the flow patterns of the Himalayan rivers with consequent impacts on the economy in terms of water availability, food security and hydropower generation. In addition, while rising population, changing life style and increasing pace of economic development are all raising demands for fresh water, the global change impacts can be all the more severe in the scenario of dipping availability of this resource. Hence understanding the impacts of global change on snow and ice dynamics in the Himalayas and their hydrological consequences has become urgent and important for a sustainable development of the region.

A regional collaborative project on "Impacts of Global Change on the Dynamics of Snow, Glaciers and Runoff over the Himalayan Mountains and Their Consequences for Highland and Downstream Regions" was initiated in the year 2008 with the support of the Asia Pacific Network for Global Change Research (APN) and with active participation of the G B Pant Institute of Himalayan Environment and Development (GBPIHED) from India, the Institute for Development and Innovation (IDI) from Nepal and Global Change Impact Studies Centre (GCISC) from Pakistan. The details on the objectives of the two-year project as well as selected research strategies and methodologies to attend the objectives have been published earlier (Shrestha 2009) and some of the results of the study have also been subsequently published (Shrestha 2012).

In the high altitude snow/glacier fed basins of the Himalayan region, the cryospheric melt runoff constitutes an important component in the total runoff. Thus generating better estimates of snow and glacier melt contribution to the Himalayan river flows and the likely change in this component due to global warming are two of the major driving concerns.

2 Methodological Approaches

As the rugged topography of Himalayas hinders field collection of scientific data in particular at high altitudes and over long periods, significant knowledge and uncertainty exist for proper understanding and assessing the impacts of global change on snow and ice dynamics as well as corresponding hydrology and consequent implications for highland and downstream communities.

In order to address such knowledge gaps and uncertainties, remote sensing and Geographical Information System (GIS) are used to determine the topographical features and to assess the status of snow and glacier dynamics of the region. Likewise, Regional Climate Models (RCMs) at the basin scales are run and used after necessary calibration and validation to provide meteorological conditions for contemporary as well as future periods. The obtained meteorological profiles with necessary bias corrections are then used to run suitable hydrological models to assess the impacts of climate change on the water availability in the selected basins both in space and time. Their consequent implications for highland and downstream communities are being studied next using chosen appropriate tools.

The general approach adopted in the research is shown in Fig. 1. The spatial scale on which a GCM operates is very coarse compared to that of hydrologic process and moreover, accuracy of GCMs, in general, decreases in regards to hydrologic variables. Hence for assessing regional hydrological implications of global climate change, properly downscaled GCMs are required in order to provide fine-resolution climate parameters to drive hydrological models for making future predictions of changes in the hydrological regimes.

Of the available two approaches of downscaling GCMs, the statistical downscaling approaches though simple from the point of view of computation, have several limitations and associated uncertainties particularly in the data sparse areas (Ghosh and Misra 2010). Hence the dynamic downscaling approaches, although computation intensive, were used to derive RCMs and used for assessing hydrologic implications of global climate change. Regional Climate Models (RCMs)

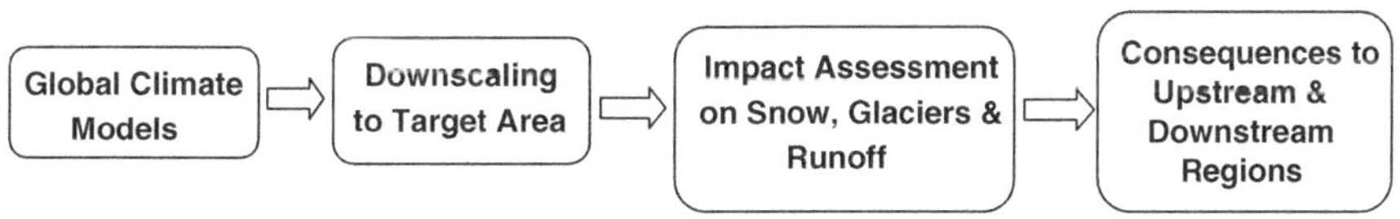

Fig. 1 Research approach

run at the basin scales are used after necessary calibration and validation to provide meteorological conditions for contemporary as well as future periods. The obtained meteorological profiles with necessary bias corrections are then used to run suitable hydrological models to assess the impacts of climate change on the water availability in the selected basins both in space and time. Their consequent implications for highland and downstream communities are next studied using chosen appropriate tools.

3 Climate Change Projections for Two Selected River Basins

3.1 *Koshi Basin*

The dynamical downscaling of different GCMs namely HadAM3P for PRECIS and ECHAM 4 for RegCM3 were taken on for the selected Koshi basin in Nepal for IPCC SRES A2 at 50 km resolutions. The projected temperatures and precipitaion for the Koshi basin in Nepal as obtained by using the two chosen RCMS are presented in Figs. 2 and 3 respectively.

While the monotonic rise in projected change in temperature, though at different slopes and intercepts, are evident in both the RCMs projections, the projected change in precipitation hardly follow any trend. Also the projected change in precipitation under Reg CM3 is much lower (the shown right hand side scale). Likewise, downscaled GCM HadCM3 for PRECIS for the Koshi basin at 25 km resolution for IPCC SRES A1B was taken on to investigate seasonal and annual temperature and precipitation changes for the basin over the three decades. As presented in Fig. 4, annual mean temperature change over the three decades show consistent increasing warming trend i.e. rise of 2.5, 3.1 and 3.5 °C from that for the baseline period for the decades of 2040s, 2050s and 2060s respectively.

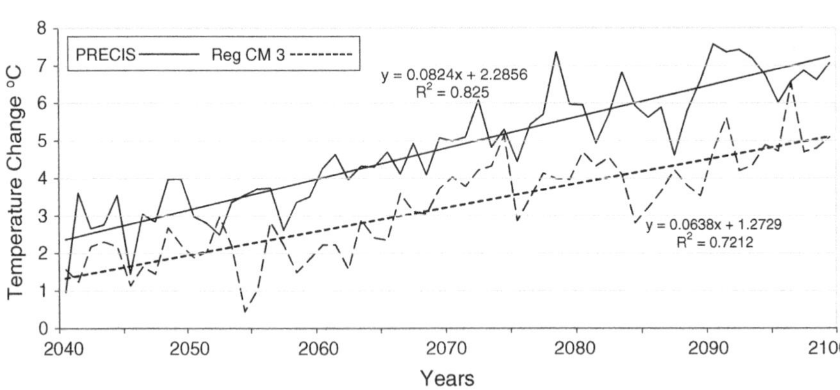

Fig. 2 RCMs projected annual mean temperature change in Koshi basin

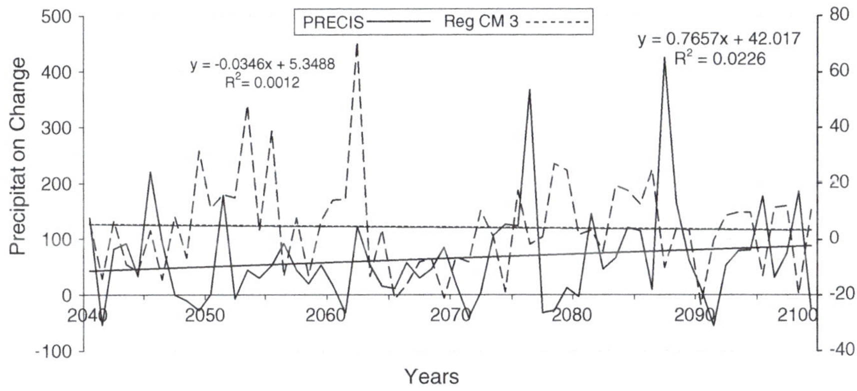

Fig. 3 RCMs projected total precipitation change in Koshi basin

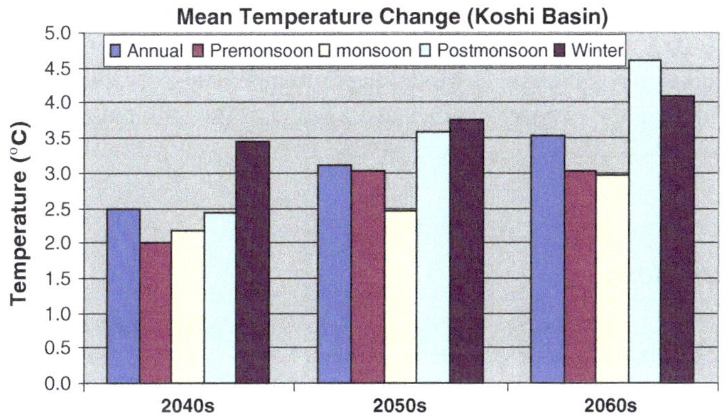

Fig. 4 Projected seasonal mean temperature changes in Koshi basin

Overall, similar trends of increasing warming for progressive decades are also found in sub-seasonal time scale; however some important exceptions are also noted. For example, projected mean seasonal rise in temperature is highest in winter for 2040s and 2050s but in 2060s post-monsoon seasonal rise is projected to become the highest. Similarly, lowest rise in temperature is projected during pre-monsoon in 2040s and during monsoon in 2050s and 2060s. This shows that amplitude of projected seasonal warming over different sub-region might vary in different decades and this could be especially true for smaller regions.

Regarding precipitation change projections as represented in Fig. 5, Koshi Basin average annual mean precipitation change shows no decadal trend; in fact, these values are very close to each other (about 7 %). Projected absolute change in precipitation is highest during monsoon for 2040s and 2050s and highest during post-monsoon in 2060s. Large seasonal decadal variability in projected

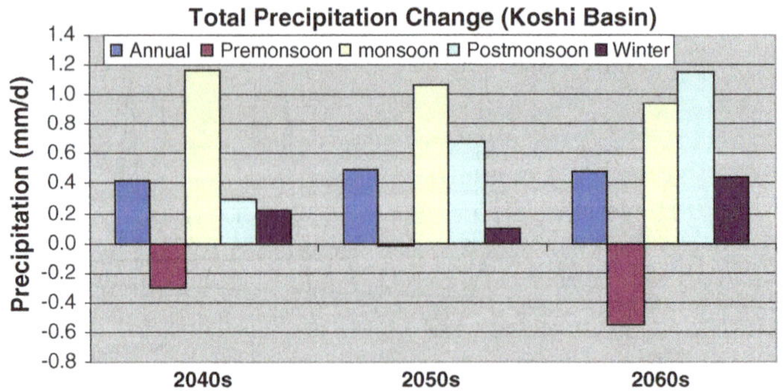

Fig. 5 Projected seasonal mean precipitation changes in Koshi basin

precipitation change is noted for all seasons except for monsoon. For example, post-monsoon projected change jumps from 0.3 mm/d (10 %) in 2040s to 1.2 mm/d (40 %) in 2060s. Therefore considerable inter-decadal variation could be expected in seasonal change. Model performance is also evaluated by comparing the computed and observed values averaged over the whole domain and model bias is thus worked out.

3.2 Hunza Basin

Likewise, climate change projections for IPCC SRES A2 using the PRECIS based on downscaling of GCM ECHAM4 for annual as well as seasonal temperature and precipitation change over Hunza basin for the various projection periods are shown in Table 1.

Over the Hunza Basin, the projected annual temperature rise of 5.48 °C by the end of the current century seems far more than the global projected temperature increase of 4.0 °C for A2. In case of precipitation about 10 % increase has been observed by the end of the current century. The seasonal analysis of temperature shows higher temperature increase in fall and winter than the spring and summer

Table 1 PRECIS climate projection for A2 over Hunza basin

		Δ Temperature °C			Δ Precipitation %		
		2020s	2050s	2080s	2020s	2050s	2080s
Annual		1.58	3.21	5.48	4.25	9.34	9.94
Seasonal	Spring (MAM)	1.39	2.97	4.25	10.22	12.37	17.15
	Summer (JJA)	1.48	2.93	5.58	−9.42	−22.76	−26.46
	Fall (SON)	1.66	3.17	6.37	−12.21	3.47	−14.70
	Winter (DJF)	1.78	3.75	5.72	13.65	23.06	33.97

temperatures. In case of precipitation, increase has been observed for spring and winter seasons throughout the century but greater increase is observed in winter than spring (i.e. 34 % with respect to base period). However, precipitation in summer and fall are found to be decreasing by up to 26 and 15 % respectively by the end of present century.

4 Snow and Glacial Melt Runoff

In the high altitude snow/glacier fed basins of the Himalayan region, the cryospheric melt runoff constitutes an important component in the total runoff. Thus generating better estimates of snow and glacier melt contribution to the Himalayan river flows and the likely change in this component due to global warming are two of the major driving concerns. Various methods and models have been applied to study the impacts of climate change on the snow melt, glacial melt and total runoff in the Himalayan rivers (Akhtar et al. 2008a, b; Immerzeel et al. 2009, 2010). Methods used in the current study and results obtained on the snow and glacier melt contributions to the total runoff in the chosen rivers are presented and discussed next.

4.1 Koshi Basin

The Koshi river, one of the major tributary of river Ganges in India, has its basin broadly cosisting of seven major sub-basins (Fig. 6). Shuttle Radar Topography Mission (SRTM) 90 m digital elevation data were used to derive the Digital Elevation Model (DEM) of the basin. This Fig. 6 Koshi Basin together with meteorological data from nearby field stations were used to derive the climate parameters at the different altitudinal bands of the selected basins. Landsat-7 ETM+ images have been used to determine extent and status of glaciers in the selected basins and as a result, 22 glacierized sub-watersheds have been identified in the Koshi basin in Nepal.

A conceptual precipitation runoff model is then used to simulate the daily runoff from identified 22 glacierized sub-watersheds. Snow and glacier melt are calculated by using surface energy balance and mass balance models. A simple empirical relation is used to define the type of precipitation. The precipitation as rainfall, snow and glacier melt in all bands are averaged and used as input to the conceptual TANK model for runoff simulation. The total melt volume comprising of glacial melt as well as snowmelt over bare ground area for the year 2002 were calculated for all smaller glaciated parts of the basins. The melt volume for seven sub-catchments of Koshi basin is then found out by summing up the ones from the associated respective individual smaller basins and is shown graphically in percentages in Fig. 7.

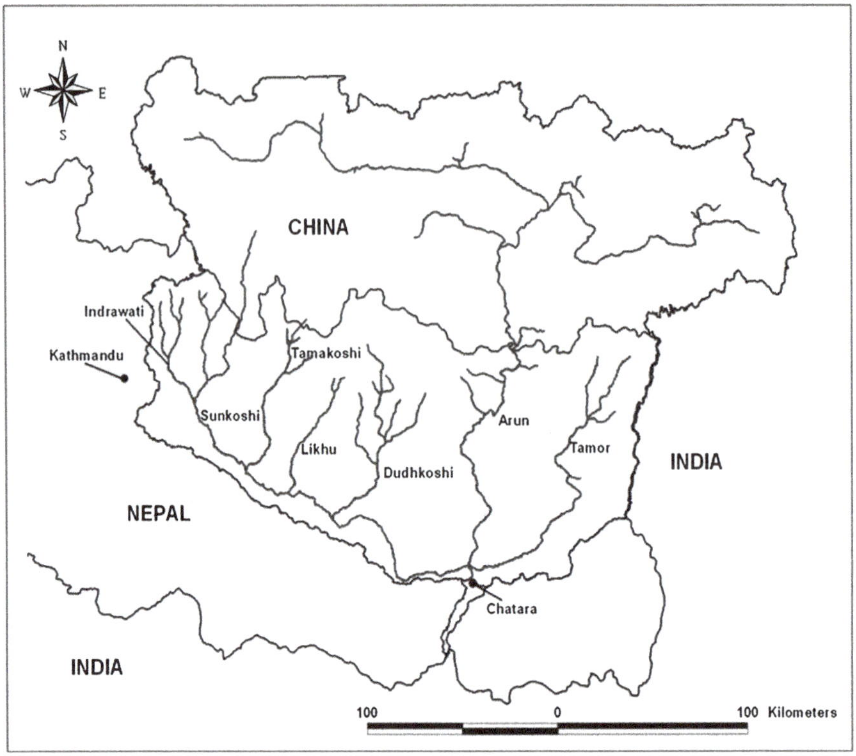

Fig. 6 Koshi basin

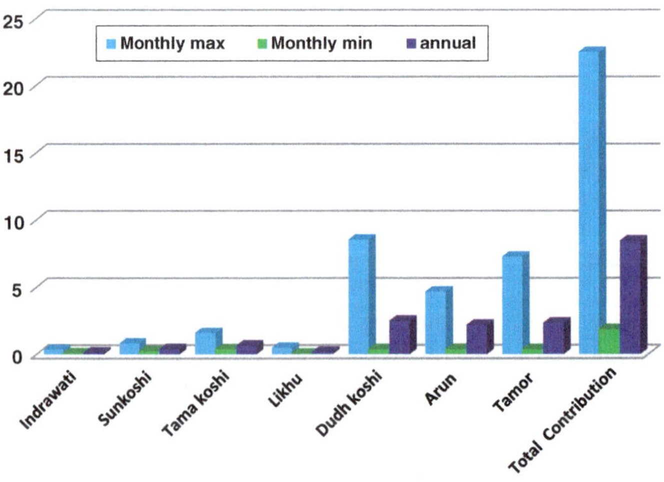

Fig. 7 Maximum and minimum percentages of snow and ice melt contributions in total flow at Chatara

The estimated runoff from the melts is then routed downstream to basin outlet in plain via seven major tributaries using a semi- distributed rainfall-runoff model namely HEC-HMS model. The analysis indicated that snow and glacier melt alone contribute maximum up to 20 % of dry season flow. The likely impact on the same in future due to climate change is being analyzed by using projected climate parameters from the RCMs. The contribution of snow and glacier melt discharge to annual flow at the lowermost downstream station ‘Chatara’ is found to be about 8.46 % with a maximum monthly contribution of 22.52 % in May and a minimum monthly contribution of 1.86 % in January. The snow and glacier melt discharge from Dudh Koshi sub-basin is found quite significant. It has maximum contribution to annual flow at Chatara (2.51 % out of total 8.46 %). Whereas glacierized watersheds of Indrawati sub-basin have minimum contribution to annual flow at Chatara (0.15 % out of total 8.46 %). Dudh Koshi, Arun and Tamor basins are three major tributaries, which share 84 % in terms of contribution of snow and glacier melt to outlet of Koshi basin at Chatara. As almost half of the Arun river basin lies in Tibet that is not included in this analysis, its cryospheric melt contribution would obviously be much more than indicated here in this analysis.

4.2 Hunza Basin

The University of British Columbia Watershed Model (UBC) was applied to study the impact of climate change in the Hunza basin. The DEM (Digital Elevation Model) for the Hunza River basin has been taken from the ASTER global dataset at horizontal resolution of 30 m. Physio-geographical characteristics such as glaciated area, forest cover, density forest canopy, impermeable area and orientation index along with the land area under each elevation zone have been calculated from the Land Cover and DEM gridded datasets from USGS. The base flow and projected flows in Hunza River including the snow melt and glacier melt components for the periods 2020s, 2050s and 2080s as obtained from the control run covering the period 1966–1990 and the scenarios for the chosen periods are shown in Table 2. Assuming the extent and the volume of glaciers in the basin remaining constant, the calculated figures clearly indicates a monotonous rise in the total runoff in the river and turning more than double of the flow during the base period. While the snow melt component is found to be not so prominent and it virtually drops to insignificance in the period 2080s, glacier melt is found to be the dominant component in the total flow throughout the period and with its contribution of about 82 % in the control period and rising continually to about 90 % in the 2080s period. Although the contribution of glacier melt in the total runoff is highly significant, the results should however be used with caution in view of the limited input data that goes into the UBC model.

Table 2 Control and scenarios for base flow and projected flows in Hunza river including the snow melt and glacier melt components

		Base			FI (2020s)			F2 (2050s)			F3 (2080s)		
		Flow	Snow	Glacier	Flow	Snow	Glacier	Flow	Snow	Glacier	Flow	Snow	Glacier
		(Cumecs)			(Cumecs)			(Cumecs)			(Cumecs)		
Annual average		292.6	6.0	243.7	371.7	6.0	314.5	473.0	6.0	407.9	654.0	4.4	588.3
Seasonal	Spring	62.7	18.3	29.0	83.8	17.9	47.7	148.3	20.0	99.1	195.8	16.9	145.4
	Summer	933.2	5.1	856.4	1,162.7	5.6	1,065.8	1,411.9	3.8	1,309.1	1,845.9	0.5	1,757.6
	Fall	154.5	0.6	89.2	217.0	0.5	144.4	304.3	0.3	223.3	539.9	0.1	449.5
	Winter	20.2	0.0	0.0	23.4	0.0	0.0	27.6	0.0	0.0	34.3	0.1	0.7

5 Conclusion and Way Forward

The Himalayan glaciers and snow together with the base flow from the high mountains contribute important components of flows rendering the rivers perennial in nature. As water resources are inextricably linked with climate and other anthropogenic changes, global change is quite likely to alter significantly the flow patterns of the Himalayan rivers with consequent impacts on the economy in terms of water availability, food security and hydropower generation. Investigations on global change impact assessment on Himalayan hydrology involves the evaluation of changes in the snow and glacier dynamics as well as runoff resulting from changes in climatic, hydrological and local anthropogenic parameters. Such an ongoing investigation carried out in a few selected watersheds in the Himalayan region under a regional collaborative effort has clearly revealed a significant hydrological impact of the ongoing global change on the chosen Himalayan watersheds.

Hydrological models forced by climate parameters from the RCMs are being calibrated and validated and then applied to make future predictions of changes in the hydrological regimes of the above selected river basins. The consequences of the projected changes in the hydrological regimes in terms of food security, hydropower development and upstream-downstream linkages are likewise being studied. Policy implications of the envisaged changes are then explored.

Acknowledgment The research presented here has been carried out as a part of the regional collaborative project funded by the Asia Pacific Network for Global Change Research (APN) with the support of the US National Science Foundation (US NSF) and the assistance of the Global Change SysTem for Analysis Research and Training (START). Their cooperation, support and assistance are gratefully acknowledged. Grateful thanks are also due to the Country-Coordinators and researchers from the collaborating partner institutions in this project namely the G B Pant Institute of Himalayan Environment and Development (GBPIHED) from India, the Institute for Development and Innovation (IDI) from Nepal and Global Change Impact Studies Centre (GCISC) from Pakistan. Special thanks are due to Dr. Lok Man S. Palni, and Dr. Rajesh Joshi from GBPIHED, India; Dr. Madan Lal Shrestha, Prof. Dr. Narendra Shakya and Mr. Jagadish Karmacharya from Nepal; and Mr. Ghazanfar Ali and Mr. Shabeh ul Hasson from GCISC, Pakistan for their valuable cooperation and contributions.

References

Akhtar M, Ahmad N, Booij MJ (2008a) The impact of climate change on the water resources of Hindu-Kush-Karakorum- Himalaya region under different glacier coverage scenarios. J Hydrol 355:148–163

Akhtar M, Ahmad N, Booij MJ (2008b) Use of regional climate model simulations as input for hydrological models for the Hindu-Kush–Karakorum–Himalaya region. Hydrol Earth Syst Sci Dis 5:865–902

Bajracharya SR, Shrestha B (eds) (2011) The status of glaciers in the Hindu Kush-Himalayan region. International Centre for Integrated Mountain Development, Kathmandu

Ghosh S, Misra C (2010) Assessing hydrological impacts of climate change: modeling techniques and challenges. Open Hydrol J 4:115–121

Gurung DR, Kulkarni AV, Giriraj A, Aung KS, Shrestha B, Srinivasan J (2011) Changes in seasonal snow cover in Hindu Kush-Himalayan region. Cryosphere Discuss 5:755–777

Immerzeel WW, Droogers P, de Jong SM, Bierkens MFP (2009) Large-scale monitoring of snow cover and runoff simulation in Himalayan river basins using remote sensing. Remote Sens Environ 113:40–49

Immerzeel WW, van Beek LPH, Bierkens MFP (2010) Climate change will affect the asian water towers. Science 328:1382–1385

IPCC (Intergovernmental Panel on Climate Change) (2007) Climate change 2007: synthesis report. In: Core Writing Team, Pachauri RK, Reisinger A (eds) Contribution of Working Groups I, II and III to the Fourth Assessment Report of the Intergovernmental Panel on Climate Change. IPCC, Geneva

IPCC (Intergovernmental Panel on Climate Change) (2010) Statement on the melting of Himalayan glaciers. http://www.ipcc.ch/publications_and_data/publications_and_data_reports.htm

Shrestha KL (2009) Impacts of global change on the dynamics of snow, glaciers and runoff over the Himalayan Mountains and their consequences for highland and downstream regions. Mountain Research Initiative (MRI) Newsletter 2(3):6–9

Shrestha KL (2012) Impacts of global change on the dynamics of snow, glaciers and runoff over the Himalayan Mountains and their consequences for highland and downstream regions. Asia Pacific Network for Global Change Research (APN) Newsletter 18(3):20–25

Identification of Glacial Lake and the Potentially Dangerous Glacial Lake in the Himalayan Basin

Sanjay K. Jain, Anil K. Lohani and Raj D. Singh

Abstract Glaciers have perennially been the source of fresh water for more than 1.3 billion of people in the Indian subcontinent. The climatic change and variability in recent decades has made considerable impacts on the glacier lifecycle in the Himalayan region. Warmer climates of the past have resulted in glacier retreat and the formation of glacial lakes in many mountain ranges. The lakes, located at the snout of the glacier, are mainly dammed by the lateral or end moraine, where there is a high tendency of breaching. Sudden discharge of large volumes of water with debris from these lakes potentially causes Glacial Lake Outburst Floods (GLOFs) in valleys downstream. A number of hydroelectric (HE) projects in India are being planned in the Himalayan regions. It has become necessary for the project planners and designers to account for the GLOF also along with the design flood for deciding the spillway capacity of projects located in similar hydro-meteorological regions. In order to assess the possible hazards from such lakes it is therefore essential to have a systematic inventory of all such lakes formed at the high altitudes. The hazardous lakes, however, are situated in remote areas and are very difficult to monitor through ground surveys due to rugged terrain and extreme climatic conditions. Monitoring of the glacial lakes and extent of GLOF impact along the downstream can be done quickly and precisely using time series satellite images and aerial photographs. To identify the individual glaciers and glacial lakes, different image enhancement techniques are useful. Besides making a temporal inventory, a regular monitoring of these lakes is also required to assess the change in their nature and aerial extent. In the present paper, identification of glacial lakes has been carried out using IRS LISSIII data. Also criteria for identification of potentially dangerous lake have been discussed.

Keywords Glaciers · GLOF · Remote sensing · Glacial lake · Himalaya

S.K. Jain (✉) · A.K. Lohani · R.D. Singh
National Institute of Hydrology, Roorkee, Uttarakhand, India
e-mail: sjain@nih.ernet.in

R. Joshi et al. (eds.), *Dynamics of Climate Change and Water Resources of Northwestern Himalaya*, Society of Earth Scientists Series,
DOI 10.1007/978-3-319-13743-8_4

1 Introduction

The Himalayan region, including the Tibetan Plateau, has shown consistent trends in overall warming during the past 100 years (Fujita et al. 1997). Various studies suggest that warming in the Himalayas has been much greater than the global average of 0.74 °C over the last 100 years (IPCC 2007). Long-term trends in the maximum, minimum and mean temperatures over the north western Himalaya during the 20th century (Bhutiyani et al. 2007) suggest a significant rise in air temperature in the north western Himalaya, with winter warming occurring at a faster rate. Global warming has remitted in large-scale retreat of glaciers throughout the world. This has led to most glaciers in the mountainous regions such as the Himalayas to recede substantially during the last century and influence stream run-off of Himalayan Rivers. The widespread glacial retreat in the Himalayas has resulted in the formation of many glacial lakes. Glacier retreat and shrinking could form dangerous moraine lakes, which can produce sudden glacier lake outburst floods (GLOFs) damaging life and property downstream over a long distance. For water resources planning and management, it is therefore essential to study and monitor the Himalayan glaciers and glacial lakes including GLOF. In this paper, glaciers, glacial lakes and glacial lake outburst flood have been presented.

Himalayan glaciers have been in a general state of recession since the 1850s (Bhambri and Bolch 2009). Existing studies of Himalayan glaciers indicate that many have exhibited an increased receding trend over the past few decades (Kulkarni et al. 2005, 2007; Chaujar 1989). Rate of glacier retreat from different parts of Himalayas are given in Table 1. Thus, regular monitoring of a large number of Himalayan glaciers is important for improving our knowledge of glacier response to climate change. Mass balance is one of the important parameters

Table 1 Rate of glacier retreat from different parts of Himalayas

Glacier and origin	Rate of retreat (m/year)
Jammu & Kashmir and Himachal Pradesh	
Barashigri, Chandan basin of Eastern Lahul	44
Tajiwas Nar, Sindh basin of Kashmir	5
Stock, Lahul	6
Gangstang, Bhara basin of western Lahul	12
Garhwal and Kumaun region, Uttarakhand	
Trishul, Nanda Devi sanctuary	10
Betharti, Nanda Devi sanctuary	8
East Kamet	5
Gangotri, Bhagirathi basin	15
Satopanth-Bhagirathi glaciers complex, Alaxnanda	12
Milan, Gori Ganga basin	13.5
Poting, Gori Ganga basin	5

Source Mukhopadhaya (2006)

which can be influenced by global warming. Geographical parameters which can influence mass balance are area-altitude distribution and orientation, since higher altitude has lower atmospheric temperature. In addition, orientation and amount of slope can influence amount of solar radiation received on the slope.

A glacial lake is defined as water mass existing in a sufficient amount and extending with a free surface in, under, beside, and/or in front of a glacier and originating from glacier activities and/or retreating processes of a glacier. The isolated lakes found in the mountains and valleys far from the glaciers may not have a glacial origin. Due to the rapid rate of ice and snow melt, possibly caused by global warming, accumulation of water in these lakes has been increasing rapidly in Himalaya. The isolated lakes above 3,500 msl are considered to be the remnants of the glacial lakes left due to the retreat of the glaciers (Campbell 2005).

The lakes located at the snout of the glacier are mainly dammed by the lateral terminal or end moraine, where there is high probability of breaching. Such lakes could be dangerous as they may hold a large quantity of water. Breaching and the instantaneous discharge of water from such lakes can cause flash floods enough to create enormous damage in the downstream areas. In order to assess the possible hazards from glacial lakes it is therefore essential to have a systematic inventory of all such lakes formed at the high altitudes. This is feasible by identifying them initially through satellite images (and aerial photographs, if available) and to assess their field setting subsequently. Besides making a temporal inventory, a regular monitoring of these lakes is also required to assess the change in their nature and aerial extent. The lakes are classified into Erosion, Valley trough, Cirque, Blocked, Moraine Dammed (Lateral Moraine and End Moraine Dammed lakes), and Supraglacial lakes (Campbell 2005).

A Glacial Lake Outburst Flood (GLOF) is created when water dammed by a glacier or a moraine is released. Some of the glacial lakes are unstable and most of them are potentially susceptible to sudden discharge of large volumes of water and debris which causes floods downstream i.e. GLOF. These floods pose severe geomorphological hazards and can wreak havoc on all manmade structures located along their path. Much of the damage created during GLOF events is associated with large amounts of debris that accompany the floodwaters. GLOF events have resulted in many deaths, as well as the destruction of houses, bridges, entire fields, forests, and roads. Unrecoverable damage to settlements and farmland can take place at large distances from the outburst source. In most of the events livelihoods are disturbed for long periods. The lakes at risk, however, are situated in remote and often inaccessible areas. When they burst, the devastation to local communities could be tremendous, while those in far away cities downstream may be unaware of the catastrophe. Many glacial lakes are known to have formed in the Himalaya in the last half century, and a number of GLOF events have been reported in the region (HKH) in the last few decades. Due to extreme hazard potential, it is necessary to take into account GLOF while planning, designing and constructing any infrastructure, especially water resources projects, as they are located on the path of flood wave and would be the prime target for catastrophe.

Since the beginning of last century the number of glacial lake outburst floods (GLOFs) increased in the Himalaya (Richardson and Reynolds 2000). Previous studies showed that the risk of lake development is highest where the glaciers have a low slope angle and a low flow velocity or are stagnant (Quincey et al. 2007; Reynolds 2000). Whether glacial lakes become dangerous depends largely on their elevation relative to the spillway over the surrounding moraine (Benn et al. 2001; Sakai et al. 2007). Triggering events for an outburst can be moraine failures induced by an earthquake, by the decrease of permafrost and increased water pressure, or a rock or snow avalanche slumping into the lake causing an overflow (Buchroithner et al. 1982; Ives 1985; Viuchard and Zimmerman 1987).

2 Identification of Glacial Lake

For glacial lakes identification from the satellite images, the image should be with minimum snow cover and cloud free. The detection of glacial lakes using multispectral imagery involves discriminating between water and other surface types. Delineating surface water can be achieved using the spectral reflectance differences. Water strongly absorbs in the near- and middle-infrared wavelengths (0.8–2.5 μm). Vegetation and soil, in contrast, have higher reflectance in the near- and middle-infrared wavelengths; hence water bodies appear dark compared to their surroundings when using these wavelengths (Pietroniro and Leconte 2000). When applying basic techniques of multispectral classification similar to those used for the normalized difference vegetation index, NDVI (Hardy and Burgan 1999), a normalized difference water index (NDWI) for lake detection was used (Huggel 1998).

$$NDWI = \frac{Green\,Band - NIR\,Band}{Green\,Band + NIR\,Band} \tag{1}$$

As a result of spectral reflection, some self shadowed areas can be misclassified as lakes. These areas can be found with the help of DEM, the DEM have to be overlaid on NDWI image. After that through manual delineation, lakes can be identified more clearly.

3 Riggering/Bursting Mechanism of Glacial Lake Outburst Floods

Different triggering mechanisms of GLOF events depend on the nature of the damming materials, the position of the lake, the volume of the water, the nature and position of the associated mother glacier, physical and topographical conditions, and other physical conditions of the surroundings.

Interaction between the above processes may strongly increase the risk of hazards. The most significant chain reaction in this context is probably the danger from ice avalanches, debris flows, rockfall or landslides reaching a lake and thus provoking a lake outburst. The mechanism of ice core-dammed and moraine-dammed lakes failure are as follows:

3.1 Mechanism of Ice Core-Dammed Lake Failure

Ice-core dammed (glacier-dammed) lakes drain mainly in two ways;

- Through or underneath the ice.
- Over the ice.

Initiation of opening within or under the ice dam (glacier) occurs in six ways.

- Floatation of the ice dam (a lake can only be drained sub-glacially if it can lift the damming ice barrier sufficiently for the water to find its way underneath).
- Pressure deformation (plastic yielding of the ice dam due to a hydrostatic pressure difference between the lake water and the adjacent less dense ice of the dam; outward progression of cracks or crevasses under shear stress due to a combination of glacier flow and high hydrostatic pressure).
- Melting of a tunnel through or under the ice.
- Drainage associated with tectonic activity.
- Water overflowing the ice dam generally along the lower margin.
- Sub-glacial melting by volcanic heat.

The bursting mechanism for ice core-dammed lakes can be highly complex and involve most or some of the above-stated hypothesis. A landslide adjacent to the lake and subsequent partial abrasion on the ice can cause the draining of ice core-moraine-dammed lakes by overtopping as the water flows over, the glacier retreats, and the lake fills rapidly.

3.2 Mechanism of Moraine-Dammed Lake Failure

Moraine dam failure and Glacial Lake Outburst Floods present one of the greatest threats to people and property in mountainous regions. In the last several centuries, many of the present moraine-dammed lakes were created by advances and retreats of valley glaciers. Moraine dams pose hazards because of their composition and location. Moraine dams are usually located downslope from steep crevassed glaciers and vertical rock slopes, and located upslope from steep canyons with easily erodible materials. Overtopping and breaching of the dam by waves generated by avalanches or glacier calving is the most common failure mechanism. The result of the dam failure is catastrophic downstream flooding. Some of the largest downstream flood peaks are produced from moraine-dam failures. Dam characteristics and failure mechanisms affect the flood peak (Costa and Schuster 1988).

3.3 Moraine Dam Formation Process

End and lateral moraines are created from material pushed and piled from glacier movement along with till released when a glacier melts and recedes. End and lateral moraines form at the glacier margin. Till is a heterogeneous accumulation of unsorted soil, rock, and other material released from the glacier ice. End moraines are formed at the farthest limit reached by a glacier. Lateral moraines are formed along the sides of a glacier. Both terminal and lateral moraines may act as dams. A moraine-dammed glacial lake is formed as the glacier recedes from the end moraine in a mountain valley. The moraine acts as an unstable dam to water melting from the glacier.

3.4 End Moraine Formation

Climate is important for the creation and stability of moraine dams. Many moraine dams and moraine-dammed lakes formed when glaciers retreated at the end of the Little Ice Age (Clague 2003; Hambrey and Alean 2004). The warming of the climate is resulting in the recession of glaciers and glacial lake formation in mountain ranges around the world. As glaciers continue to melt, more lakes are created and the size of the lakes is increasing. Failure of moraine dams is becoming increasingly frequent. Failures have occurred in the Himalayas, European Alps, North American Rockies, Andes, Tien Shan Mountains, and Pamirs (Reynolds 2006). The frequency and magnitude of moraine dam failures and Glacial Lake Outburst Floods will continue to increase with the current and continued scenario of global warming (RGSL 2003).

The unstable nature of moraine dams greatly increases the chances of dam failure. Many factors contribute to the unstable nature of the moraine dam. The moraine dam may contain stagnant glacier ice. Moraine dams are typically narrow and high reducing structural strength. Large lake volume above the moraine dam increases pressure against the dam. Limited freeboard between the crest of the moraine dam and the lake level reduces the height of waves needed to overtop the moraine dam (Reynolds 2003). Since moraine dams have steep slopes and contain loose, poorly sorted sediment, moraine dams are highly susceptible to failure (Clague 2003). Moraine dam failure may be caused from one of several triggers. Large waves caused from calving glaciers or ice or rock avalanches into the lake may overtop the moraine dam. Settlement within the dam due to voids left during deposit may reduce the freeboard or create passageways for piping to occur. Melting of stagnant glacier ice in the moraine dam may also reduce the freeboard or create passageways for piping to occur. Catastrophic glacial drainage may raise the lake level quickly and overtop the dam (RGSL 2003; Hambrey and Alean 2004). Earthquakes may also cause moraine dam failure (Clague 2003).

4 Criteria for Identification of Potentially Dangerous Glacial Lakes

The potentially dangerous lakes can be identified based on the condition of lakes, dams, associated mother glaciers, and topographic features around the lakes and glaciers. The criteria used to identify these lakes are based on field observations, processes and records of past events, geomorphological and geo-technical characteristics of the lake and surroundings, and other physical conditions. Identification was also based on the condition of lakes, dams, associated glaciers, and topographic features around the lakes and glaciers. The major criteria used for identification of potentially dangerous glacial lakes are as follows (Mool et al. 2001a, b; Mool and Bajracharya 2003; Bhagat et al. 2004; Sah et al. 2005; Roohi et al. 2005; Wu et al. 2005):

1. Large lake size and rapid growth in area
2. Activity of supra-glacial lakes at different times
3. Position of the lakes in relation to moraines and associated glacier
4. Dam condition
5. Glacier condition
6. Physical conditions of surroundings.

4.1 Dam Condition

(a) Narrow crest area, (b) No drainage outflow or outlet not well defined, (c) Steepness of slope of the moraine walls, (d) Existence and stability of ice core and/or permafrost within moraine, (e) Height of moraine, (f) Mass movement, or potential mass movement, on the inner slope and/or outer slope of moraine, (g) Breached and closed in the past and the lake refilled with water, (h) Seepage through the moraine walls.

4.2 Glacier Condition

(a) Condition of associated glacier, (b) Hanging glacier in contact with lake, (c) Large glacier area, (d) Rapid glacier retreat, (e) Debris cover on the lower glacier tongue, (f) Gradient of glacier tongue, (g) Presence of crevasses and ponds on glacier surface, (h) Toppling/collapsing of ice from the glacier front, (i) Ice blocks draining to lake.

4.3 Physical Conditions of Surroundings

(a) Potential rockfall/slide (mass movement) sites around the lake, (b) Large snow avalanche sites immediately above the lake, (c) Neo-tectonic and earthquake activities around or near the lake, (d) Climatic conditions, especially large inter-annual variations, (e) Very recent moraines of tributary glaciers that were previously part of a former glacier complex, and with multiple lakes that have developed due to glacier retreat, (f) Sudden advance of a glacier towards a lower tributary or main glacier which has a well-developed frontal lake.

Once the dangerous lake is identified, simulation of the potentially dangerous lake using hydro dynamic modeling is carried out to know the flood at different locations. In NIH some of the studies on GLOF have been carried out for basins located in Eastern Himalaya and Bhutan Himalaya. In these studies, Lake Inventory has been prepared using satellite data. The area and perimeter etc. of all the lakes located in the basin were computed. Potentially dangerous lake has been identified with the help of remote sensing data. Data required for GLOF studies have been generated in GIS. GLOF simulation has been carried out using MIKE 11 software.

5 Conclusion

The Himalayan glaciers are huge storage and very important source of fresh water. The glaciers are one of the most sensitive indicators of climate change as they grow and shrink in quick response to changing air temperature. The Himalayan ice and glaciers are gradually melting due to global temperature rise resulting to significant shrinkage in snow-covered area, retreating of glaciers at rate of tens of meters per year and formation of glacier lakes. These changes are greatly affecting runoff patterns and increasing the risks of GLOF. For assessment of total volume and stream flow pattern from different glaciers along with glacier retreat and GLOF comprehensive studies are needed.

References

Benn DI, Wiseman S, Hands K (2001) Growth and drainage of supraglacial lakes on debris-mantled Ngozumpa Glacier, Khumbu Himal Nepal. J Glaciol 47(159):626–638

Bhagat RM, Kalia V, Sood C, Mool PK, Bajracharya S (2004) Inventory of glaciers and glacial lakes and the identification of potential glacial lake outburst floods (GLOFs) affected by global warming in the mountains of the Himalayan region: Himachal Pradesh Himalaya, India. Unpublished project report, Himachal Pradesh Agricultural University, Palampur, India

Bhambri R, Bolch T (2009) Glacier mapping: a review with special reference to the Indian Himalayas. Progr Phys Geogr 33(5):672–704

Bhutiyani MR, Kale VS, Pawar NJ (2007) Long-term trends in maximum, minimum and mean annual air temperatures across the Northwestern Himalaya during the 20th century. Clim Change 85(1–2):159–177

Buchroithner MF, Jentsch G, Waniverhaus B (1982) Monitoring of recent glaciological events in the Khumbu Area (Himalaya, Nepal) by digital processing of Landsat MSS data. Rock Mech 15(4):181–197

Campbell JG (2005) Inventory of glaciers, glacial lakes and the identification of potential glacial lake outburst floods (GLOFs) affected by global warming in the Mountains of India, Pakistan and China/Tibet autonomous region, Technical report of APN project 2004-CMY-Campbell

Chaujar RK (1989) Glacial geomorphology—a case study of Chota Shigri Glacier. In: Proceedings of national meet on Himalayan glaciology, New Delhi. Ministry of Science and Technology, Government of India, 173–185

Clague JJ (2003) Catastrophic floods caused by sudden draining of lakes in high mountains. In: Martin K, Hik DS (eds) The state of ecological and earth sciences in mountain areas. Science highlights from symposium on Ecological and Earth Sciences in Mountain Areas, Banff, Alberta, Canada, p 46

Costa JE, Schuster RL (1988) The formation and failure of natural dams. Geol Soc Am Bull 100:1054–1068

Fujita K, Nakawo M, Ageta Y, Shankar K, Pokhrel AP, Yao T (1997) Basic studies for assessing the impacts of the global warming on the Himalayan Cryosphere, 1994–1996. Bull Glacier Res 15:53–58

Hardy CC, Burgan RE (1999) Evaluation of NDVI or monitoring moisture in three vegetation types of the western U.S. Photogram Eng Remote Sens 65:603–610

Huggel C (1998) Periglaziale Seen im Luft- und Satellitenbild. Diploma thesis, Department of Geography, University of Zurich

IPCC (2007) Climate Change 2007: the physical science basis. In: Solomon S, Qin D, Manning M, Chen Z, Marquis MC, Averyt K, Tignor M, Miller HL (eds) Contribution of Working Group 1 to the Fourth Assessment Report of the Intergovernmental Panel on Climate Change. Intergovernmental Panel on Climate Change, Cambridge

Kulkarni AV, Rathore BP, Mahajan S, Mathur P (2005) Alarming retreat of Parbati glacier, Beas basin, Himachal Pradesh. Curr Sci 88(11):1844–1850

Kulkarni AV, Bhauguna IM, Rathore BP, Singh SK, Randhawa SS, Sood RK (2007) Glacial retreat in Himalaya using Indian remote sensing satellite data. Curr Sci 92(1):69–74

Mool PK, Bajracharya SR (2003) Inventory of glaciers, glacial lakes and the identification of potential glacial lake outburst floods (GLOFs) affected by global warming in the mountains of Himalayan region: Tista Basin, Sikkim Himalaya, India. Unpublished project report prepared for Asia Pacific Network for Global Change Research (APN), Japan and International Centre for Integrated Mountain Development (ICIMOD), Kathmandu

Mool PK, Bajracharya SR, Joshi SP (2001a) Inventory of glaciers, glacial lakes, and glacial lake outburst floods: Monitoring and early warning systems in the Hindu Kush-Himalayan regions—Nepal. Technical report, International Centre for Integrated Mountain Development (ICIMOD), Kathmandu

Mool PK, Wangda D, Bajracharya SR, Kunzang K, Gurung DR, Joshi SP (2001b) Inventory of glaciers, glacial lakes, and glacial lake outburst floods: Monitoring and early warning systems in the Hindu Kush-Himalayan region—Bhutan. Technical report, International centre for integrated mountain development (ICIMOD), Kathmandu

Mukhopadhaya SC (2006) Glaciers and water resources appraisal of Himalaya with special reference to Central Himalaya, Uttranchal. In: Rawat MMS (ed) Resources appraisal technology application, and environmental challenges in Central Himalaya, vol 1. Transmedia Publication, Srinagar, pp 106–119

Paul F (2001) Evaluation of different methods for glacier mapping using Landsat TM. Workshop on Land, Ice and Snow, Dresden/FRG. Proceedings EARSeLe, vol 1. pp 239–245

Pietroniro A, Leconte R (2000) A review of Canadian remote sensing applications in hydrology, 1995–1999. Hydrol Process 14:1641–1666

Quincey DJ, Richardson SD, Luckman A, Lucas RM, Reynolds JM, Hambrey MJ, Glasser NF (2007) Early recognition of glacial lake hazards in the Himalaya using remote sensing datasets. Global Planet Change 56(1–2):137–152

Reynolds JM (2000) On the formation of supraglacial lakes on debris-covered glaciers. In: Nakawo M, Raymond CF, Fountain A (eds) Debris-covered glaciers, vol 264. International Association of Hydrologic Sciences Publication, pp 153–161

Reynolds JM (2006) Role of geophysics in glacial hazard assessment. First Break 24:61–66

Reynolds Geo-Sciences (2003) Development of glacial hazard and risk minimization protocols in rural environments: guidelines for the management of glacial hazards and risks. Project report R7816. Mold, United Kingdom: Reynolds Geo-Sciences Ltd. Flintshire

Richardson SD, Reynolds JM (2000) An overview of glacial hazards in the Himalayas. Quatern Int 65(6):31–47

Roohi R, Ashraf R, Naz R, Hussain SA, Chaudhry MH (2005) Inventory of glaciers and glacial lakes outburst floods (GLOFs) affected by global warming in the mountains of Himalayan region, Indus Basin, Pakistan Himalaya. Report prepared for International Centre for Integrated Mountain Development (ICIMOD), Kathmandu

Sah M, Philip G, Mool PK, Bajracharya S, Shrestha B (2005) Inventory of glaciers and glacial lakes and the identification of potential glacial lake outburst floods (GLOFs). Unpublished report, Wadia Institute of Himalayan Geology, Dehradun, India

Sakai A, Saito M, Nishimura K, Yamada T, Izuka Y, Harado K, Kobayashi S, Fujita K, Gurung CB (2007) Topographical survey of end moraine dead ice area at Imja Glacial Lake in 2001 and 2002. Bull Glaciol Res 24:29–36

Vuichard D, Zimmermann M (1987) The 1985 catastrophic drainage of a moraine-dammed lake, Khumbu Himal, Nepal-cause and consequences. Mt Res Dev 7:91–110

Wu L, Che T, Jin R, Li Xin, Gong T, Xie Y, Mool PK, Bajracharya S, Shrestha B, Joshi SP (2005) Inventory of glaciers, glacial lakes and the identification of potential glacial lake outburst floods (GLOFs) affected by global warming in the mountains of Himalayan region: Pumqu, Rongxer, Poiqu, Zangbuqin, Jilongcangbu, Majiacangbu, Daoliqu, and Jiazhagangge basins, Tibet Autonomous Region, People's Republic of China. Unpublished project report, prepared for Asia Pacific Network for Global Change Research (APN), Japan and International Centre for Integrated Mountain Development (ICIMOD), Kathmandu

Assessment and Simulation of Glacial Lake Outburst Floods for Dhauliganga Basin in Northwestern Himalayan Region

Anil K. Lohani, Sanjay K. Jain and Raj D. Singh

Abstract Glacial lake outburst flood (GLOF) is created when water dammed by a glacier or a moraine is released. Some of the glacial lakes are unstable and most of them are potentially susceptible to sudden discharge of large volumes of water and debris which causes floods downstream i.e., GLOF. Many glacial lakes are known to have formed in the Himalaya in the last half century, and a number of GLOF events have been reported in the region in the last few decades. Due to extreme hazard potential of GLOF events, it is necessary to take into account GLOF while planning, designing and constructing any infrastructure, especially water resources projects, as they are located on the path of glacial lake outburst flood wave and would be the prime target in case of GLOF. GLOF modeling may be carried out by either scaled physical hydraulic models or mathematical simulation using computer. A modern tool to deal with this problem is the mathematical model, which is most cost effective and reasonably solves the governing flow equations of continuity and momentum by computer simulation. Mathematical modeling of dam breach floods can be carried out by either one dimensional analysis or two dimensional analysis. In one dimensional analysis, the information about the magnitude of flood, i.e., discharge and water levels, variation of these with time and velocity of flow through breach can be obtained in the direction of flow. In the case of two dimensional analyses, the additional information about the inundated area, variation of surface elevation and velocities in two dimensions can also be assessed. In the present paper, methodology for simulation of glacier lake outburst floods has been discussed. Further, the discussed methodology has been demonstrated in Dhauliganga basin located in Northwestern Himalayan region. For the present case, MIKE-11 model of Danish Hydraulic Institute has been selected in this study because of its modeling accuracy for the slopes steeper than 0.01 and wide acceptability in more than 40 countries.

Keywords Glacial lake outburst flood (GLOF) · Simulation · GLOF modeling · Himalayan region · MIKE-11

A.K. Lohani (✉) · S.K. Jain · R.D. Singh
National Institute of Hydrology, Roorkee, Uttarakhand, India
e-mail: aklnih@gmail.com

R. Joshi et al. (eds.), *Dynamics of Climate Change and Water Resources of Northwestern Himalaya*, Society of Earth Scientists Series,
DOI 10.1007/978-3-319-13743-8_5

1 Introduction

The Indian Himalayan region is home to a numerous glaciers, which is nature's renewable storehouse of fresh water. According to a report by the International Centre for Integrated Mountain Development (ICIMOD) there are 15,000 glaciers and 9,000 glacial lakes in the Himalaya (Mool 2005). The Himalayan region is intrinsically linked to global atmospheric circulation, hydrological cycle, biodiversity and water resources and play a crucial role in shaping and influencing the Indian environmental conditions. Due to global climatic change observed during the first half of the twentieth century, tremendous impact on the high mountainous glacial environment has been observed. It has been reported by the various researchers that many of the big glaciers melted rapidly and gave birth to the origin of a large number of glacier lakes.

The accelerated global warming may accelerate the retreating/melting of glaciers in the Himalayan region and thus lead to accumulation of increasing amounts of water in mountain-top lakes. As glaciers retreat, glacial lakes located at the snout of the glacier are normally formed behind moraine or ice 'dams'. Due to the inherent instability of such 'dams', the potential of sudden outbursts/breaches is extremely high. Such lakes could be dangerous as they may hold a large quantity of water and such outbursts can lead to a very high discharge of millions of cubic meters of water and debris in a few hours. Breaching and the instantaneous discharge of water from such lakes can cause flash floods in the downstream areas and it may lead to serious damage to life, property, agriculture, livestock, forests, eco-systems and engineering and water resources structures such as Hydro power projects. ICIMOD, in collaboration with partners in different countries, has prepared an inventory of glaciers, glacial lakes and identified potentially dangerous glacial lakes and potential sites for GLOF (ICIMOD 2010). Researchers have identified a number of glacial lakes and found that they are expanding as a consequence of climate change and glacier thinning (Reynolds 2000; Ageta et al. 2000; Benn et al. 2000). A number of studies related to assessment of potential of glacial lake outburst (Watanabe et al. 1994), estimation of the rate of expansion of a glacial lake (Sakai et al. 2000), automatic detection method of the lake surface using a normalized difference water index (NDWI) (Bolch et al. 2008). Some studies have asserted the expansion rate as posing a risk of a glacial lake outburst flood (GLOF) (Quincey et al. 2007), determining the trigger and proof strength of the damming moraine is more important.

2 Study Area

Dhauliganga basin up to Dhakar (outlet) is selected as the study area for the present study (Fig. 1). Dhauliganga River is the northernmost right bank tributary of river Kali and lies entirely in the Pithoragarh district of Uttarakhand state of India. The Dhauliganga basin is bounded between latitude $29^{\circ}55'$–$30^{\circ}35'$ N and longitude $80^{\circ}15'$–$80^{\circ}45'$ E. This is a glacial and snowfed river and originates in the

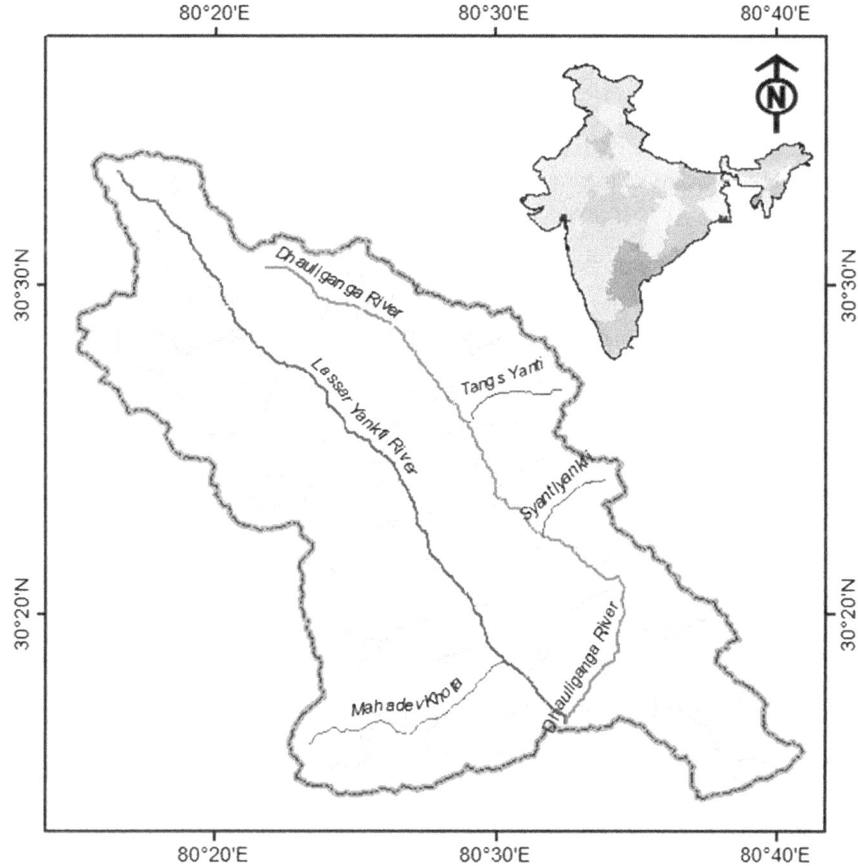

Fig. 1 Study area

lesser Himalayas from snow peaks at an elevation of 5,160 m and flows down as a small stream in a narrow valley, generally in NE–SE direction over a length of 85 km before joining the Kali river at an elevation of approximate 1,100 m. The average bed slope of the Dhauliganga River is approximately 1 in 20 and can be termed as a fast flowing ferocious river. The river valley is located in high mountain ranges on both banks over most of its stretch. The river brings down a considerable amount of sediment load particularly during snow-melt and flood season.

3 Methodology

In order to identify the potentially dangerous glacial lakes, a digital database of glaciers and glacial lakes is necessary. Different image enhancement techniques are useful for the identification of the individual glaciers and glacial lakes.

However, complemented by the visual interpretation method (visual pattern recognition), with the knowledge and experience of the terrain conditions, glacier and glacial lake inventories and monitoring can be done. Using the combinations of different spectral band in false colour composite (FCC) and in individual spectral bands, glaciers and glacial lakes can be identified and studied using the knowledge of image interpretation keys such as, colour, tone, texture, pattern, association, shape, shadow, etc. Different colour composite images highlight different land-cover features.

The lake water in colour composite images ranges in appearance from light blue to blue to black. In the case of frozen lakes, it appears white. Sizes are generally small, having circular, semi-circular, or elongated shapes with very fine texture and are generally associated with glaciers in the case of high lying areas, or rivers in the case of low lying areas. The ERDAS imagine 9.3 and Arc GIS 9.3 have been used for the processing of satellite data and GIS analysis. In the present study, map of Dhauliganga basin has been prepared using ASTER DEM. The total area is 4,782.9 km^2 and the elevation values ranges from a minimum of 1,238 to maximum of 7,785 m with in the study area. The present study has utilized mainly digital remote sensing data from the IRS-1D/P6 LISS-III sensor. When applying basic techniques of multispectral classification similar to those used for the normalized difference vegetation index, NDVI (Hardy and Burgan 1999), a normalized difference water index (NDWI) for lake detection was used (Huggel et al. 2002).

$$NDWI = \frac{B_{NIR} - B_{blue}}{B_{NIR} + B_{blue}} \tag{1}$$

As a result of spectral reflection, some self shadowed areas are misclassified as lakes. These areas have been found with the help of DEM, the DEM was over-laid on NDWI image. It could thus be assured that only lake appeared as black spots. After that, manual delineation of lakes has been carried out. Lake boundaries were digitized using ERDAS Imagine vector module tools. The digitized polygons have been assigned polygon ID's. The area of the lake was calculated using the digital techniques by counting the number of pixels falling inside the water body polygon. The geographic latitude and longitude of the centre of the lake has been computed using attributes information of the polygon later. There may be a possibility of some lakes which are snow covered and can not be fully distinguishable in the satellite data. The lakes or water bodies which are partly snow covered or fully snow covered and could not be distinguishable are not reported.

3.1 Identification of Potentially Dangerous Glacial Lakes

The criteria for identifying potentially dangerous glacial lakes are based on field observations, processes and records of past events, geo-morphological and geo-technical characteristics of the lake and surroundings, and other physical conditions. Criteria such as area/volume of lake, breaching evidences, condition of lake and its surrounding etc. have been investigated. For this purpose remote sensing data in conjunction with Google Earth have been used. It was observed that no lake is located in ablation zone of glacier and not connected with mother glacier. There were no breaching evidences and area of lake was not expanded much in the past. There is no lake which is vulnerable from GLOF point of view, however, breaching of the biggest lake has been considered for demonstrating the simulation of GLOF in the present study. Arc-GIS and ERDAS Imagine software were used to delineate cross-sections of the stream.

For this purpose the vector layer of the stream and the buffer lines along the stream on the both side of stream at the distance of 1 km were created. The stream was divided at the distance of 5 km from lake side and the cross-section layer was created. ERDAS Imagine Software was used to overlay DEM of basin and vector layer of cross-section. The Spatial Profile Viewer in ERDAS allows to visualize the reflectance spectrum of a polyline of data file values in a single band of data (one-dimensional mode) or in many bands (perspective three-dimensional mode). This is being used to create a height cross-section profile along a route. This helps in interpreting changes in elevation along a planned route and in identifying sections of the route which are particularly steep or flat. Inquire cursor of ERDAS Imagine was used to extract the elevation values at each pixel.

3.2 Glacial Lake Outburst Flood (GLOF) Simulation

Glacial lake outburst floods increase to peak flow than gradually or abruptly decrease to normal levels once the water source is exhausted. Outburst flood peak flow is directly related to lake volume, dam height and width, dam material composition, failure mechanism, downstream topography, and sediment availability. In order to get the maximum GLOF peak at any location, the breaching of moraine dams of above glacial lakes have to be considered along with channel routing. In order to get the maximum GLOF peak at the outlet, the out bursting or breaching of glacial lakes have been considered along with channel routing. Arc-GIS software was used to delineate cross-section area along the stream. For this purpose the vector layer of the stream and the buffer lines along the stream on the both side of stream at the distance of 1 km were created. The stream was divided at the distance of 1 km from lake side and the cross-section layer was created. Points were created along cross section line and the lines were converted into point file. Then this file was crossed with DEM and elevation at all the points were obtained.

There is no estimate available for the estimation of volume of glacial lakes in Himalaya from their water spread areas. However, Huggel et al. (2002) has given some estimates for glacial lakes in Swiss Alps. In the absence of information on the volume of potentially dangerous glacial lakes, it is considered appropriate to use the same relationships developed for the lakes in Swiss Alps for estimating the water volume for the lakes in this area.

The empirical relations used by Huggel et al. (2002) are:

$$\text{lake volume } V = 0.104\ A^{1.42} \tag{2}$$

where V is the lake volume in m^3 A is the lake area in m^2.

$$\text{lake depth } D = 0.104\ A^{0.42} \tag{3}$$

where D is the lake depth in m.

Due to possible out bursting or breaching of lake existing in the study area may result a surge of flood at the outlet. Therefore, in order to estimate the maximum GLOF peak at the outlet, the out bursting or breaching of lake have to be considered along with channel routing. This study demonstrates the simulation of the flood hydrograph just upstream of selected outlet in the event of breaching of largest glacial lakes. Thus, the estimation of GLOF (Sharma et al. 2009) provides the flood hydrograph of discharge from the dam breach and maximum water level at different locations of the river downstream of the dam due to propagation of flood waves along with their time of occurrence.

The essence of dam break modelling is hydrodynamic modelling, which involves finding solution of two partial differential equations originally derived by Barre De Saint Venant in 1871. These equations are:

Continuity equation:

$$(\partial Q/\partial X) + \partial (A + A_0)/\partial t - q = 0 \tag{4}$$

Momentum equation:

$$(\partial Q/\partial t) + \left\{\partial\left(Q^2/A\right)/\partial X\right\} + g A\left((\partial h/\partial X) + S_f + S_C\right) = 0 \tag{5}$$

where, Q = discharge; A = active flow area; A_0 = inactive storage area; h = water surface elevation; q = lateral flow; x = distance along waterway; t = time; S_f = friction slope; S_c = expansion contraction slope and g = gravitational acceleration.

The mathematical models which approximately solve the governing flow equations of continuity and momentum by computer simulation are the cost effective modern tool for the dambreak analysis. MIKE-11 and DAMBRK computer programs have been developed in the recent past; however these computer programs are dependent on certain inputs regarding the geometric and temporal characteristics of the dam breach. The state-of-art in estimating these breach characteristics is not as advanced as that of the computer programs, and therefore these are the limiting factors in this analysis.

4 Results

4.1 GLOF Simulation for Biggest Lake

There is no lake which is vulnerable from GLOF point of view, however, in the present study, largest lake has been considered to carry out GLOF study to demonstrate impact of GLOF at downstream areas. This lake is at a distance of 25 km from outlet. The Dhauliganga River from glacial lake location down to the outlet has been represented in the model by a number of cross sections at an interval of 1 km, developed from DEM. The locations of cross sections are shown in Fig. 2 and cross sections at an interval of 1 km downstream of lake are shown in Fig. 3. The altitude of the Lake is 4,339 m. The surface area of the lake and altitude are given in Table 1. The volume of the lake is calculated using Eq. 2 taking lake area as 1,25,000 m^2 and it comes out to be 1.79 mm^3. The depth of the lake is computed using Eq. 3 and it comes to be 15 m. The breaching starts at top of the dam and continues up to the breach invert level. The mode of failure for glacial lake outburst considered is overtopping.

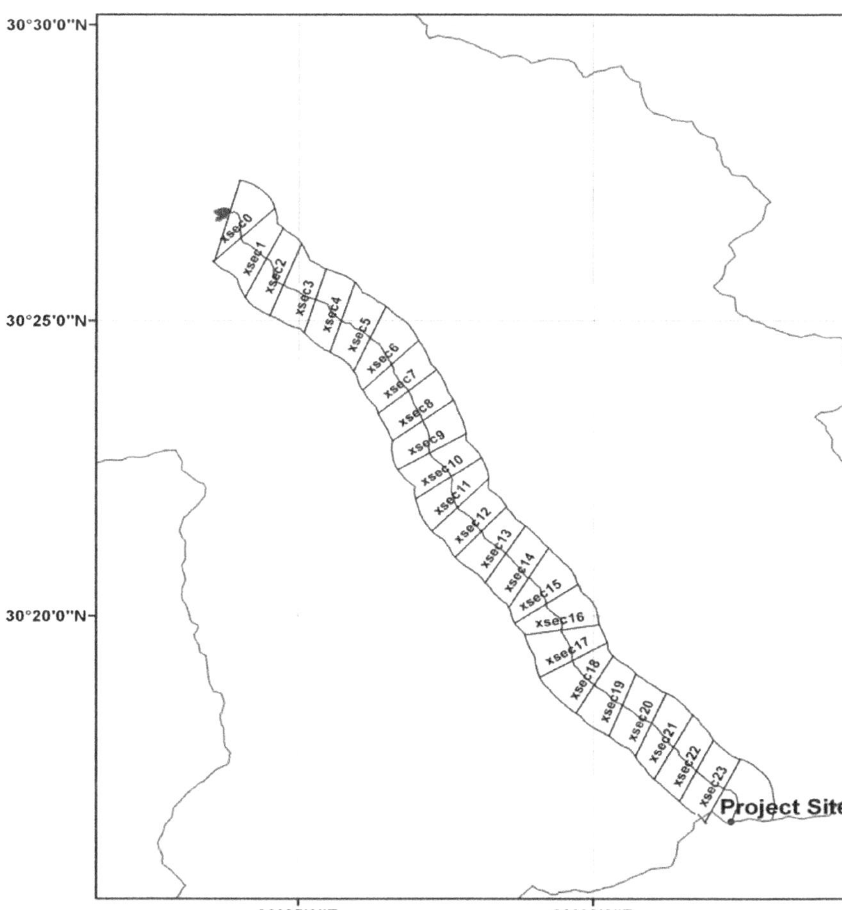

Fig. 2 Location of cross sections at 1 km interval downstream of lake

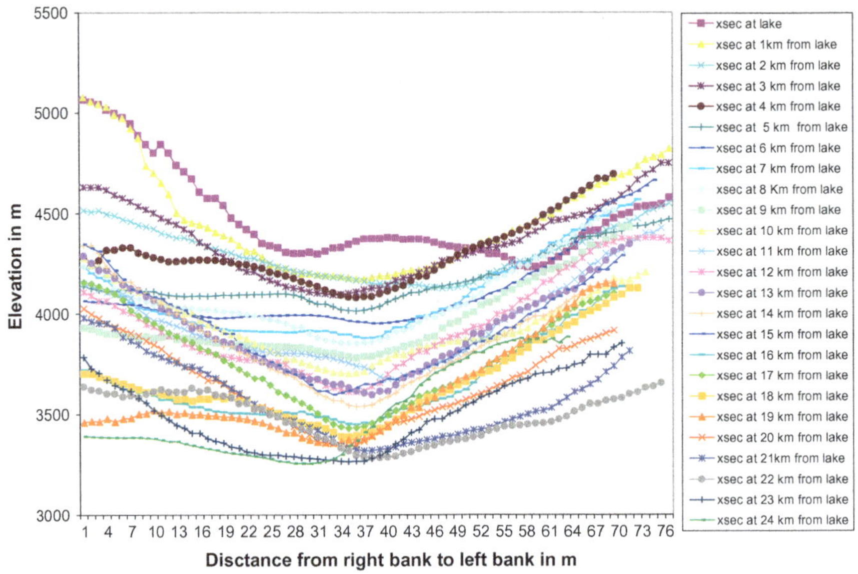

Fig. 3 Cross sections at 1 km interval downstream of lake

Table 1 Elevation-area relationship of glacial lakes

Glacial lake (volume 1.79 m^3)	
Elevation (m)	Surface area (m^2)
4,323.000	115,000.00
4,324.000	115,000.00
4,339.000	125,000.000

The breach invert level should be taken as two-third to three-fourth of the height of dam below its top level, therefore, breach depth has been taken as 12 m which is approximately three fourth of 15 m. For average breach width, equation given by FERC 1987 has been used. As per this equation, breach width comes out to be in the range of 30–60 m. Using this range of width and going through the satellite image of the lake, the breach width has been taken as 50 m. The breach development time has been computed using equation given by FERC 1987 and it is from 0.1 to 1 h. Mool et al. (2001) has given breach development time as 1,000–2,000 s. Considering the size of the lake, it is appropriate to assume the breach development time of 30 min to get the maximum GLOF peak. The side slope has been taken as 1H:1V. The Manning's roughness coefficient has been taken as 0.05 considering the boulder beds and hilly terrain of Himalayan.

The 100 year return period flood value available at outlet is distributed along the study reach on the basis of the catchment areas of the contributing tributaries. There are mainly four streams which are meeting with the stream coming from lake up to outlet. The distributed flood values have been taken as lateral inflows of magnitude corresponding to 100 year flood with their impingement locations in the Dhuliganga River, as given in Table 2. MIKE 11 software was applied for generation of flood hydrograph for breach width of 50 m. These flood hydrographs at just downstream of the lake and near outlet are shown in Fig. 4. The total flood peak and its travel time from GLOF site is given in Table 3. It can be seen from the Table 2, that the GLOF peak is 808.66 m^3/s and the same get mitigated to

Table 2 Lateral inflow considered corresponding to 100 year return period flood

Sl. no.	Location	100 year flood (m^3/s)	Adopted lateral inflow (m^3/s)	Area (km^2)
1	Upstream stream from lake	215.11	215.1143	138.87
2	7.85 km downstream from lake	325.37	110.2602	71.18
3	10.25 km downstream from lake	370.31	44.93747	29.01
4.	16.75 km downstream from lake	474.62	104.3119	67.34
5	20.21 km downstream from lake	602.63	128.0121	82.64
6	25.21 km downstream from lake	1,082.00	479.4259	309.5

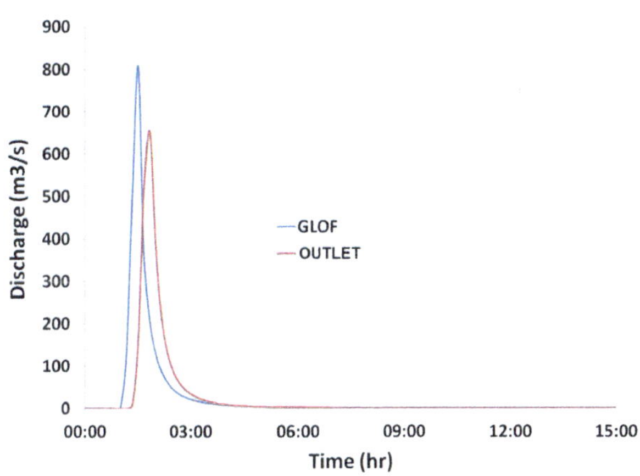

Fig. 4 GLOF hydrograph at the outlet

Table 3 Flood peak due to glacial lake outburst for breach width of 50 m

Location	Distance (km) from glacial lake	Breach width 50 m	
		GLOF peak (m^3/s)	Travel time (hr-min)
Just d/s of lake	0	808.66	00-00
Outlet	25	652.87	0-20

652.9 m^3/s at the outlet for breach width of 50 m. The time of travel of flood peak from the lake site to outlet is about 20 min.

5 Conclusion

Glacial lake outburst flow modelling process is an approximation of a physical phenomenon through which the physical phenomenon and its effects can be studied for water resources structure design and flood management. In GLOF modeling, assumptions are mainly associated with the breach parameters, especially, breach width and breach depth, which has impact on flood peak and arrival times. In general, glacial lake bursting mechanism and formation of breach in glacial lakes are not fully understood. Furthermore, the high velocity associated with GLOF can cause significant scour of channels associated with bed as well as bank erosion. Change in the channel cross section due to GLOF is neglected due to limitations in modelling such a complicated physical process. Generally, GLOF creates a large amount of transported debris and this may be accumulate at constricted cross sections, where it acts as a temporary dam and partially or completely restricts the flow, resulting variation in flood peak arrival time. This aspect has also been neglected due to limitations in modelling of such a complicated physical process. These limitations have an effect mainly on the conservative side. Even with the assumptions and limitations outlined above, hydrodynamic modelling serves very useful purpose, as it provides reasonable estimate of glacial lake outburst flood, thus enabling the appropriate estimation of design flood.

The integration of visual and digital image analysis with a Geographic Information System (GIS) can provide very useful tools for the study of glacial lakes, and Glacial Lake Outburst Floods (GLOFs). MIKE 11 software was applied for generation of flood hydrograph for breach width of 50 m. A flood of 100 year return period has been considered in addition to GLOF. The 100 return flood has been taken as 1,082 m^3/s at outlet. The distance of outlet from biggest lake is 25 km. GLOF peak is 808.66 m^3/s and the same get mitigated to 652.87 m^3/s at the outlet for breach width of 50 m. The time of travel of flood peak from the lake site to outlet is about 20 min.

References

Ageta Y, Iwata S, Yabuki H, Naito N, Sakai A, Narama C, Karma T (2000) Expansion of glacier lakes in recent decades in the Bhutan himalayas. In: Debris-covered glaciers, proceedings of the workshop, vol 264, Seattle, USA, International Association of Hydrological Sciences Publication, pp 165–175

Benn D, Wiseman S, Warren C (2000) Rapid growth of a Supraglacial lake, Ngozumpa Glacier, Khumbu Himal, Nepal. In: Debris-covered glaciers, proceedings of the workshop, vol 264, Seattle, USA, International Association of Hydrological Sciences Publication, pp 177–186

Bolch T, Buchroithner MF, Peters J, Baessler M, Bajracharya S (2008) Identification of glacier motion and potentially dangerous glacial lakes in the Mt. Everest region, Nepal using spaceborne imagery. Nat Hazards Earth Syst Sci 8:1329–1340

Hardy CC, Burgan RE (1999) Evaluation of NDVI for monitoring moisture in three vegetation types of the western U.S. Photogram Eng Remote Sens 65:603–610

Huggel C, Kääb A, Haeberli W, Teysseire P, Paul F (2002) Remote sensing based assessment of hazards from glacier lake outbursts: a case study in the Swiss Alps. Can Geotech J 39:316–330

ICIMOD (2010) Formation of glacial lakes in the Hindu Kush-Himalayas and GLOF risk assessment. In: Ives JD, Shrestha RB, Mool PK (eds) International Centre for Integrated Mountain Development, Kathmandu

Mool PK (2005) Monitoring of glaciers and glacial lakes from 1970s to 2000 in Poiqu Basin, Tibet autonomous region, PR China. Unpublished report of international centre for integrated mountain development (ICIMOD), Kathmandu and Cold and Arid Regions Environmental and Engineering Research Institute, Gansu, China

Mool PK, Bajracharya SR, Joshi SP (2001) Inventory of glaciers, glacial lakes and glacial lake outburst floods, Nepal. International Centre for Integrated Mountain Development, Kathmandu, Nepal

Quincey DJ, Richardson SD, Luckman A, Lucas RM, Reynolds JM, Hambrey MJ, Glasser NF (2007) Early recognition of glacial lake hazards in the himalaya using remote sensing datasets. Glob Planet Change 56(1–2):137–152

Reynolds JM (2000) On the formation of Supraglacial lakes on debris-covered glaciers. Int Assoc Hydrol Sci Publ 264:153–161

Sakai A, Chikita K, Yamada T (2000) Expansion of a moraine-damned glacial lake, Tsho Rolpa, in Rolwaling Himal, Nepal Himalaya. Limnol Oceanogr 45(6):1401–1408

Sharma D, Bhagat S, Perumal M, Rai NN (2009) Study of glacial lake outburst flood for punatsangchhu hydro electric project, Bhutan. In: National symposium on climate change and water resources in India, Indian Association of Hydrologists (IAH), Roorkee

Watanabe T, Ives JD, Hammond JE (1994) Rapid growth of a glacial lake in Khumbu Himal, Nepal: prospects for a catastrophic flood. Mt Res Dev 14:329–340

A Model Study of Dokriani Glacier, Garhwal Himalaya, India

Argha Banerjee and R. Shankar

Abstract One dimensional flowline models have been widely used to understand the past glaciations of various glaciers, and also to predict their future behaviour. We report results from a simplified flowline model simulation of Dokriani Glacier, Uttarakhand. The measured mass balance reported earlier for the period of 1991–2000, is used as input. The geometry of the glacier is also taken into account in a simple way. We assume that the glacier was close to a steady state around 1962 and that the equilibrium line altitude is moving up with a uniform rate since then. Our simulated glacier length data matches with available length record of this glacier for the period of 1962–2000 quite well. We discuss the model predictions for the future behaviour of this glacier for various possible warming rates.

Keywords Debris covered glacier · Glacier modelling · Flowline model · Glacier geometry · Dokriani glacier

1 Introduction

Understanding the future behaviour of various Himalayan glaciers is crucial because of their importance as hydrological resources. This necessitates modelling studies of these glacier systems. A general glaciological model can be tuned to describe a particular glacier using the measured values of a few basic field observables. This model can then be used to describe the past records, and to predict the future trends taking input from climate model predictions. This framework has been successfully employed to study various Himalayan glaciers, using various types of models (Naito et al. 2000; Adhikari and Huybrechts 2009). In this paper, we have used one dimensional flowline model (Oerlemans 1989, 2001) to study Dokriani glacier, Uttarakhand, India. Dokriani glacier is a partially debris covered glacier. A full

A. Banerjee (✉) · R. Shankar
The Institute of Mathematical Sciences, Chennai, India
e-mail: arghab@imsc.res.in

R. Joshi et al. (eds.), *Dynamics of Climate Change and Water Resources of Northwestern Himalaya*, Society of Earth Scientists Series,
DOI 10.1007/978-3-319-13743-8_6

description of this glacier thus requires a coupled flowline model of debris and ice (Naito et al. 2000). But we take an alternative approach, accounting for the effect of debris through an effective specific mass balance function in a simpler setting of a one dimensional flowline model for the ice flow. We use available geometrical data and mass balance data from the study of Dobhal et al. (2007) to calibrate this model. We assume a uniform warming rate during the period 1962–2002 that coincides with the warming rate observed during 1992–2002. Result from this model successfully describes the length variation of this glacier for the period of 1962–2002 (Dobhal et al. 2007). The model is also used to predict the future behaviour of this glacier under various possible uniform warming rates for the period of 2002–2042.

2 Study Area

The present study has been carried out on Dokriani glacier (30°49′–30°52′ N and 78°47′–78°51′ E) located in the Bhagirathi river basin in Garhwal Himalaya, of Uttarakhand. The glacier was originally mapped in 1962–1963 and was remapped in 1995 by the Survey of India. Dokriani glacier is one of the well-developed, medium sized (7.0 km^2) valley glaciers of Gangotri group of glaciers in the Garhwal Himalaya. It originates at an elevation of 6,000 m amsl and is formed by two cirque glaciers. The glacier follows NNW direction for about 2 km before it turns towards WSW and terminates at an altitude of 3,886 m (Fig. 1). The length

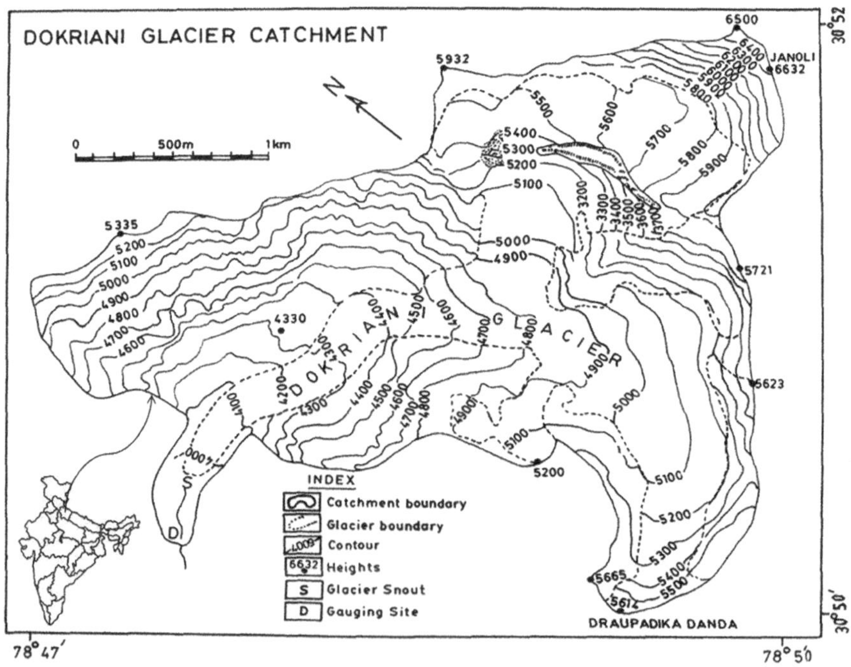

Fig. 1 Location map of Dokriani glacier (After: Thayyen et al. 1999)

of the glacier is 5.5 km with a width varying from 0.08 to 2.5 km. The total catchment area is 15.7 km^2, with the glacier ice covering an area of 7.0 km^2 whereas thickness of glacier ice varies from 25 to 120 m between snout and accumulation zone. Lower reaches of the ablation zone are covered by debris, while the lateral moraines exist all the way along the glacier up to the accumulation zone (Gergan et al. 1999).

3 Methodology

3.1 Flowline Model

One-dimensional flowline model (Oerlemans 1989, 2001) provides a simple way to describe a valley glacier (Oerlemans 1997; Wallinga and Van der Wal 1998; Adhikari and Huybrechts 2009). The model dynamics is determined by local conservation of ice,

$$\partial tH = 1/w\,\partial x(HUw) + B \tag{1}$$

where, H is the thickness of ice, w is the cross-sectional area, U is the cross-section averaged ice velocity along the flowline, B is mass balance function, x is the longitudinal coordinate, and t denotes time. The mean flow velocity can be parameterised by the following relation,

$$U = fd(gH\,\partial x(H+z))^3 H + fs(gH\,\partial x(H+z))^3/H \tag{2}$$

here fd and fs are two parameters controlling contributions of ice deformation and sliding to the ice ow, and z(x) defines the bedrock geometry. Given z(x) and B(x; t), Eq. (1) provides a complete description of any particular glacier.

3.2 A Simple Model of Dokriani Glacier

Dokriani glacier is a partially debris covered Himalayan valley glacier, situated in the Bhagirathi basin and is about 5.5 km long. Dobhal et al. (2007) have measured its specific mass balance curve as a function of altitude and the snout retreat for the period of 1992–2000. The extent of this glacier in 1962 is also known from maps published by Survey of India. We model this glacier using a simplified mass balance function (Fig. 2a) that approximates the measured specific balance rates for the period 1992–2000 (Fig. 2). Compared to the mass balance curve of a typical debris free glacier, where melting increases monotonically at lower elevations, the mass balance curve of Dokriani glacier nearly saturates to a constant ablation rate in the lower part of the ablation zone. This is a typical feature of the mass balance function of debris covered glaciers where insulation provided by a thick debris layer inhibits melting in the lower ablation zone.

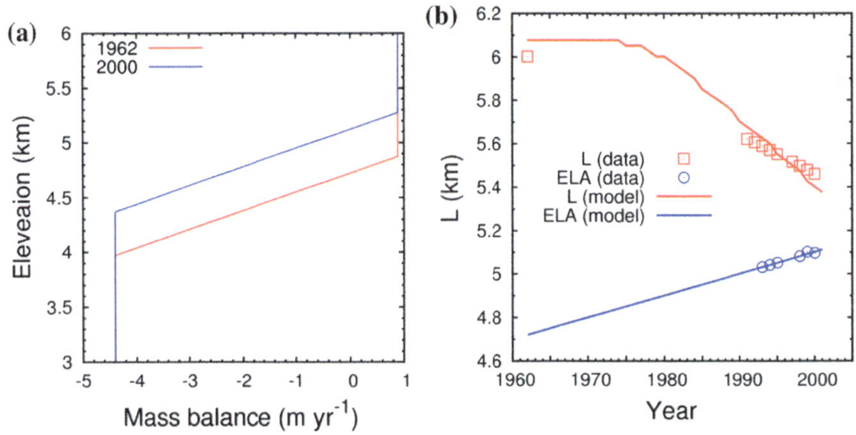

Fig. 2 **a** Approximate mass balance function used to model Dokriani glacier (Based on field data from Dobhal et al. 2007). Between 1962 and 2000, the whole curve is assumed to be moving up at an uniform rate. **b** Comparison of available length data and model result for the period 1962–2000. The assumed variation of ELA and corresponding available data are also shown

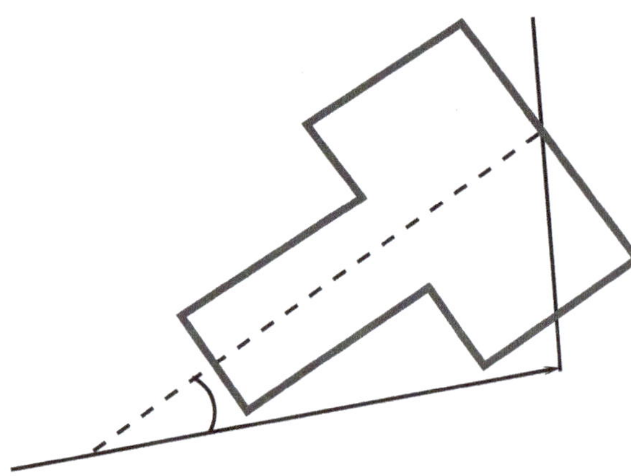

Fig. 3 simplified geometry used to model Dokriani glacier

To model the effect of a warming climate, the equilibrium line altitude (ELA) is moved up uniformly with time, keeping the shape of the mass balance curve the same. We assume that the glacier was in a steady state around 1962 with ELA at 4,870 m and that the ELA has been moving up with a constant rate of 10 m/year. These values are obtained from the best fit linear extrapolation of the ELA data (Dobhal et al. 2007) for the period of 1992–2000. A uniform bedrock slope with 0.4 is used and the area above the elevation of 5,000 m is taken to be 3.3 times larger than that below this elevation. This simplified geometry is shown in Fig. 3.

We have used fd and fs values that are smaller than those used by Oerlemans (2001) by a factor of 1/4. This has been done to match the observed average ice thickness values of 50 m (Gergan et al. 1999) for this glacier following Adhikari and Huybrechts (2009). Also note that we follow the numerical scheme described by Oerlemans (2001) to solve the flowline equation and our temporal and spatial discretization steps are 0.00025 year and 25 m respectively.

4 Model Results

4.1 The Past Data

As shown in Fig. 2b, our simple model does very well in reproducing the length variation data for the period of 1962–2002. This shows that despite our simplifying assumptions the model does capture the behaviour of the real glacier reasonably well. One interesting feature of the modelled response is that while the warming has been assumed to start in 1962 in our model, the glacier length has remained remarkably steady for a period of about 15 years after that subsequently the glacier has undergone steady retreat. This surprising behaviour is related to the presence of debris cover in the lower ablation zone of this glacier. Here, the insulating effects of thickening debris layer causes ablation to saturate to a minimum value, which otherwise would have increased towards lower elevations. As a result, the initial response of a debris covered glacier to warming is thinning of the glacier while the length remains steady. After this initial delay the length also starts shrinking.

4.2 Future Trends

Using the same model we have explored the future behaviour of this glacier under various possible warming rates as shown in Fig. 4. As expected higher warming rates leads to faster retreat and smaller warming rates slows the rate of length retreat. Notably, even as the warming rate is assumed to change after 2002, it takes about 20 years for this change to affect the steady retreat rate. This is true even in the case where temperature becomes steady after 2002. This, again, can be related to the presence of the debris cover which modifies the nature of the response properties of the glaciers because of its insulating effects.

Fig. 4 Model prediction for retreat of Dokriani glacier for an assumed ELA movement rate of 0 m/year (*green*), 5 m/year (*dark blue*), 10 m/year (*purple*) and 15 m/year (*sky blue*) during the period of 2002–2042

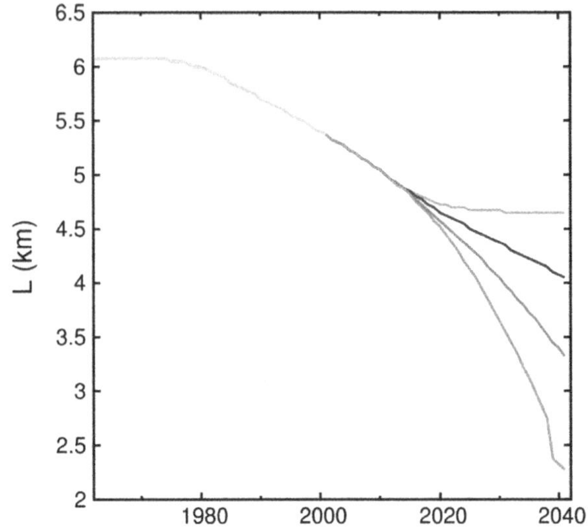

5 Conclusion

We have performed a flowline line model simulation of Dokriani glacier using a simpled geometry and an eective mass balance function. Our model results describe the retreat data for this glacier for the period of 1962–2002 reasonably well. We also simulate the future behaviour of this glacier under various climatic conditions for the period of 2002–2042. Interestingly the response of this glacier to change in prevailing steady climatic condition is characterised by a time scale of about 15 years during which the length remains unaffected by the change. This is typical of a debris covered glacier where initial climatic response is through thinning and the shrinking of the glacier length starts after a characteristic delay.

References

Adhikari S, Huybrechts P (2009) Numerical modelling of historical front variations and the 21st century evolution of glacier AX010, Nepal Himalaya. Ann Glaciol 50(52):27–34

Dobhal DP, Gergan JT, Thayyen RJ (2007) Mass balance and snout recession measurements (1991–2000) of Dokriani glacier, Garhwal Himalaya, India. Technical document in hydrology No. 80, UNESCO, Paris, pp 53–63

Gergan JT, Dobhal DP, Kaushik R (1999) Ground penetration radar ice thickness measurement of dokriani bamak (glacier) Garhwal Himalaya. Curr Sci 77(1):169–174

Naito N, Nakawo M, Kadota T, Raymond CF (2000) Numerical simulation of recent shrinkage of Khumbu glacier, Nepal Himalayas. In: Debris covered glaciers, vol 264. International Association of Hydrologic Sciences Publication, pp 245–254

Oerlemans J (1989) Glacier fluctuations and climatic change. Kluwer Academic Publishers, Dordrecht, pp 353–371

Oerlemans J (1997) A flowline model for Nigardsbreen, Norway: projection of future glacier length based on dynamic calibration with the historic record. Ann Glaciol 24:382–389

Oerlemans J (2001) Glaciers and climate change. A Balkema Publishers, Rotterdam

Thayyen RJ, Gergan JT, Dobhal DP (1999) Particle size characteristics of suspended sediments and subglacial hydrology of Dokriani glacier, Garhwal Himalaya India. Hydrol Sci J 44(1):47–61

Wallinga J, Van der Wal RSW (1998) Sensitivity of Rhonegletscher, Switzerland, to climate change: experiments with a one-dimensional flowline model. J Glaciol 44(147):383–393

Part II
Assessment of Climate Change Patterns

Critical Evaluation and Assessment of Average Annual Precipitation in The Indus, The Ganges and The Brahmaputra Basins, Northern India

Abul A. Khan, Naresh C. Pant, Anuj Goswami, Ravish Lal and Rajesh Joshi

Abstract Three major river basins of India which include the Indus, the Ganges and the Brahmaputra contribute more than 50 % of the river discharge of the country. Widely varying average annual precipitations have been reported for these basins. The average annual precipitation is a basic input data for any developmental planning. Low density of rain gauge stations especially in mountainous area, extreme variation in altitudes and large size of these basins forces adaption of remote sensed data for estimation of average annual precipitation. In the present study, 11-years (2000–2010) Tropical Rain Measurement Mission (TRMM) generated radar precipitation raw data has been used for estimating the annual precipitation. The results indicate 434, 1,094 and 2,143 mm annual precipitation for the Indus, the Ganges and the Brahmaputra basins respectively. The contoured distribution of precipitation indicates the orographic control as the primary factor on the summer monsoon precipitation in the Ganges and the Brahmaputra basins. Indus basin behaves independent of the Indian summer monsoon.

Keywords Average annual precipitation · Indus · Ganges · Brahmaputra · TRMM · Monsoon

---

A.A. Khan · N.C. Pant (✉) · R. Lal
Department of Geology, University of Delhi, Delhi, India
e-mail: pantnc@gmail.com

A. Goswami
Manipal Institute of Technology, Manipal, Karnataka, India

R. Joshi
G.B. Pant Institute of Himalayan Environment and Development, Almora, Uttarakhand, India

R. Joshi et al. (eds.), *Dynamics of Climate Change and Water Resources of Northwestern Himalaya*, Society of Earth Scientists Series,
DOI 10.1007/978-3-319-13743-8_7

1 Introduction

Precipitation constitutes an important component of the hydrological cycle especially in tropical regions. In domains of highly variable relief, such as in case of the Himalayas and the adjacent regions, the precipitation exhibits a complex relationship wherein it is orographically controlled but at the same time has a first order influence on the landscape development (Anders et al. 2006). Influence of precipitation on large terrigenous transport and its societal impact (e.g. energy generation) cannot be over emphasized. Most significantly the direct influence of precipitation on human water consumption (potable water, agriculture, etc.), especially in the high population density areas, measurement of short and long term precipitation is a basic and essential input for sustenance and every facet of developmental planning. The drainage basins of the Indus, the Ganges and the Brahmaputra not only cover ~50 % area of India (Fig. 1) but also control the hydrological requirements for nearly half of India's population. The geomorphic diversity of these basins is reflected in the variation of altitudes from 4,500 to 6,000 m in the headwaters to the sea level (Fig. 2).

On account of this snow and glacial melt contributes variably to the total discharges of these basins besides surface run-off and ground water making the

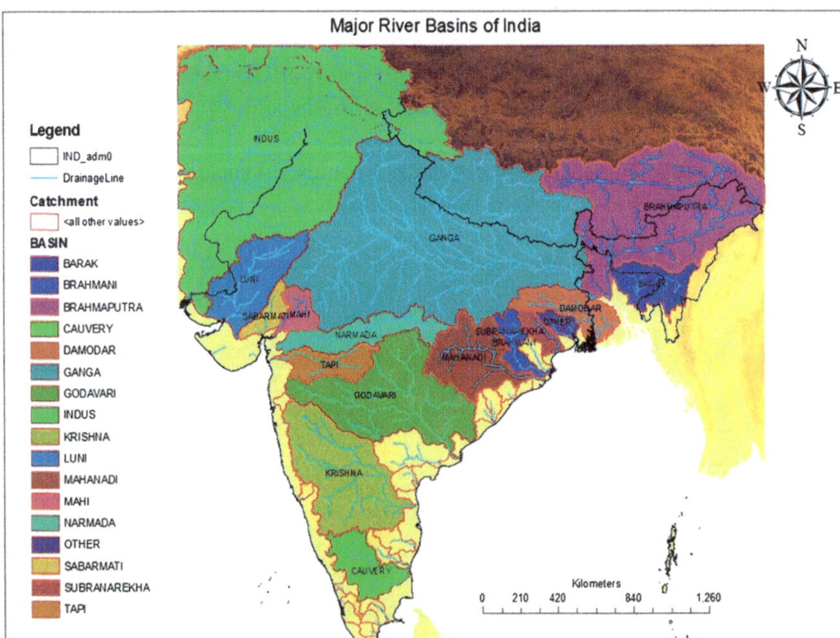

Fig. 1 Map showing drainage basins of India. Note drainage basin area of the Indus, the Ganges and the Brahmaputra basins which covers more than 50 % of the area of India. The map is drawn from the SRTM raw data

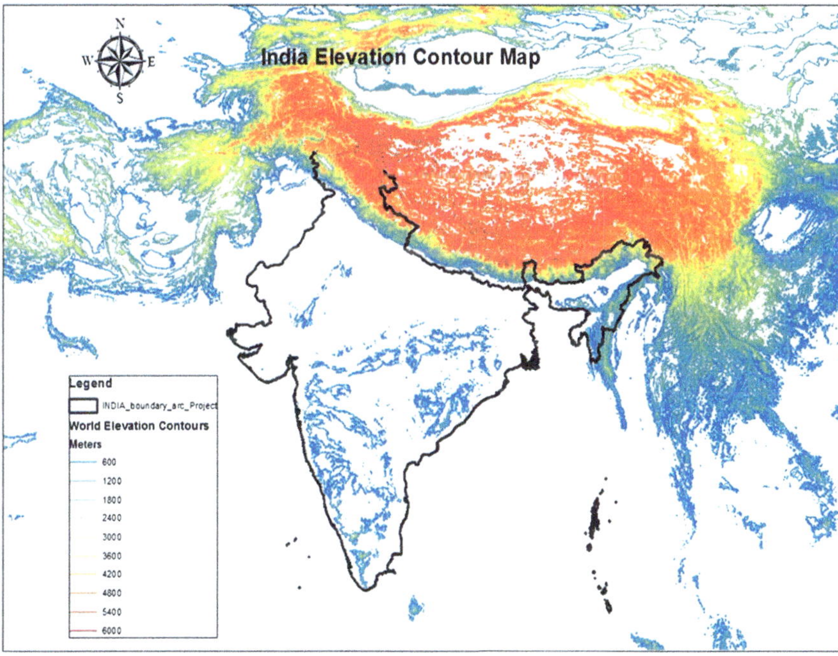

Fig. 2 Map of India showing variation of altitudes. Note extreme altitude variations in the Indus, the Ganges and the Brahmaputra basins. The map is drawn from the SRTM raw data

assessment of the hydrological cycle a complex phenomenon. The relative contributions of each of these sources are debatable (Maurya et al. 2011; Immerzeel et al. 2010) but surface run-off controlled by both liquid and solid precipitation remains the major contributor. Thus, the significance of long and short term precipitation measurement cannot be overemphasized in hydrological budgeting, agricultural and hydroelectric planning and natural disaster (e.g. floods and earth mass movements) management. Currently available average annual precipitation measurements for these three basins are summarized in Table 1. The southwest Indian summer monsoon controls the majority of precipitation (Bookhagen et al. 2005; Bookhagen and Burbank 2006) which falls as snow in the high elevation and rainfall at low elevation and adjacent flatlands (Bookhagen et al. 2010).

It is clearly seen that there is a wide variability in the average annual precipitation estimates (Table 1). The reported average annual precipitation for the Indus basin varies from 423 mm (Immerzeel et al. 2010) to 1,035 mm (Jain and Kumar 2012) while for the Brahmaputra basin it varies from 1,097 (Immerzeel et al. 2010) to more than 2,500 mm (Jain and Kumar 2012). The estimated average annual precipitation for the Ganges basin shows less variation and is reported to be ~1,000 mm (Table 1). The present paper attempts to estimate long term averaged annual precipitation for these three basins using radar data with a view to provide a better estimate of annual rainfall in these three basins.

Table 1 Reported average annual basin precipitation (mm) of three major basins

Indus	Ganges	Brahmaputra	References
423	1,035	10,711	Immerzeel et al. (2010)
1,051.2	1,097.1	2,589.2	Jain and Kumar (2012)

2 Methodology

Considering the high variability of tropical rainfall, its significance as contributor of nearly two third of the total global rainfall and satellite based remote sensing as the only reliable rainfall data on regional scale, a Tropical Rainfall Measuring Mission (TRMM) satellite was jointly launched by the US and Japan in 1997. This mission uses microwave and visible/infrared sensors for estimation of precipitation. The TRMM 3B42-V7 provides estimates of daily precipitation at about 27×27 km^2 pixel ($0.25° \times 0.25°$) between north latitude 50° and south latitude 50° by combining precipitation radar (PR) and microwave imager (TMI) data (http://disc.sci.gsfc.nasa.gov/precipitation/documentation/TRMM_README/TRMM_2B31_readme.shtm). We used raw rainfall data for a 11-year period (2000–2010) for rectangular grids containing the three basins. The method of estimation is illustrated for one basin. For Ganges basin this grid was between north latitudes 22.375° and 32.625° and east longitude 73.375° and 89.875° (Fig. 3).

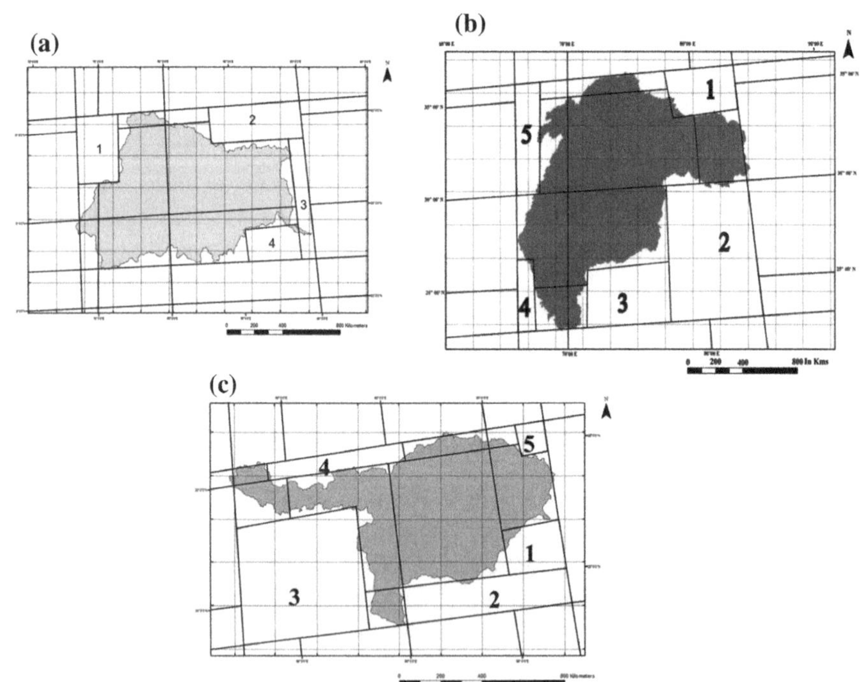

Fig. 3 The outline map of **a** Ganges basin, **b** Indus basin, and **c** Brahmaputra basin recovered from the SRTM data. *Blocks marked* indicate the domains from where the radar precipitation data has been excluded

Within this grid a geo-referenced outline of the basin was plotted using SRTM data and excess area (extra-basinal area and the rainfall within these) approximated and excluded through four rectangular domains guided by the pixel size (marked as 1, 2, 3 and 4 in Fig. 3a) to arrive at the total rainfall approximation for the entire basin. Total number of raw TRMM pixels for the Ganges basin rectangular grid work out to be 2,278 (34 × 67). For extra basin domain 1 (Fig. 3) calculated over 27.125°–30.625°N and 73.375°–76.625°E there are 210 pixels (15 × 14); for domain 2 calculated over 28.75°–30.625°N latitude and 83.625°–89.875°E longitude there are 182 pixels (7 × 26); for domain 3 calculated over 22.375°–28.875°N latitude and 89.125°–89.875°E longitude there are 108 pixels (27 × 4) and for domain 4 calculated over 22.375°–24.375°N latitude and 86.125°–88.875°E longitude there are 108 pixels (9 × 12) with each pixel separated by 0.25° along latitude as well as longitude. The pixel for the excess, non-basinal area (210 + 182 + 108 + 108 = 608) are than summed and subtracted for the rectangular area which contains the basin (2,278 pixels) resulting in a total of 1,670 pixels (2,278 − 608 = 1,670) for the Ganges basin. The same method is applied for the Indus and the Brahmaputra basins for which the grid containing the basins as well as the domains estimated for exclusion for basin wise rainfall estimation are shown in Fig. 3b, c.

2.1 Average Annual Rainfall Estimation

2.1.1 Indus Basin

Indus basin is unique amongst the three basins in many ways. It is the largest basin amongst the three with the basin area in excess of 1 million km^2 (Table 2). After its origination from China, the Indus river follows a northwest course through India and thereafter it takes a southwest course in Pakistan for most part of its journey to the Arabian Sea (Fig. 1). The two recently available estimates of the annual precipitation for the Indus basin show a wide variation from 423 mm (Immerzeel et al. 2010) to 1,051 mm (Jain and Kumar 2012) though the data from the later work appeared to have been derived from a part of the Indus basin only (Table 2). Though the total glaciated area of this basin is intermediate between the three basins, the glacial melt component in the total discharge nearly four times the other two basins at ~45 % (Table 2). The basin area as recovered from the SRTM images has been estimated as 1,273,259 km^2 which is higher than that reported by other researchers (Table 2).

Representative data from the 11-year annual precipitation for northern India and the adjacent regions is shown in the annualized TRMM maps in Fig. 4. Even considering the draught year (2002) and excess rainfall year (2003) it is quite evident that a large part of the Indus basin is relatively rain deficient. Contour diagrams for the annual precipitation estimated from the radar data for three representative years have been plotted on the georeferenced outline of the Indus basin

Table 2 Compilation of annual average precipitation of the Indus basin

Total basin area (km^2)	Glaciated area (km^2)	Number of glaciers	Annual basin precipitation in (mm)	Glaciated area in (%)	Glacier melt (%)	Mean discharge (m^3/s)	Estimated ice reserves (km^3)	Total population	References
1,005,786			423	2.2	26			209,619	Immerzeel et al. (2010)
1,081,718					44.8	5,533		178,483	Jianchu et al. (2009), IHP/HWRP report, IUCN/IWMI et al. (2002), Mi and Xie (2002), Chalise and Khanal (2001), Merz (2004)
321,289			1,097.1						Jain and Kumar (2012)
	21,193	18,495					2,696.05		Bajracharya and Shrestha (2011), ICIMOD

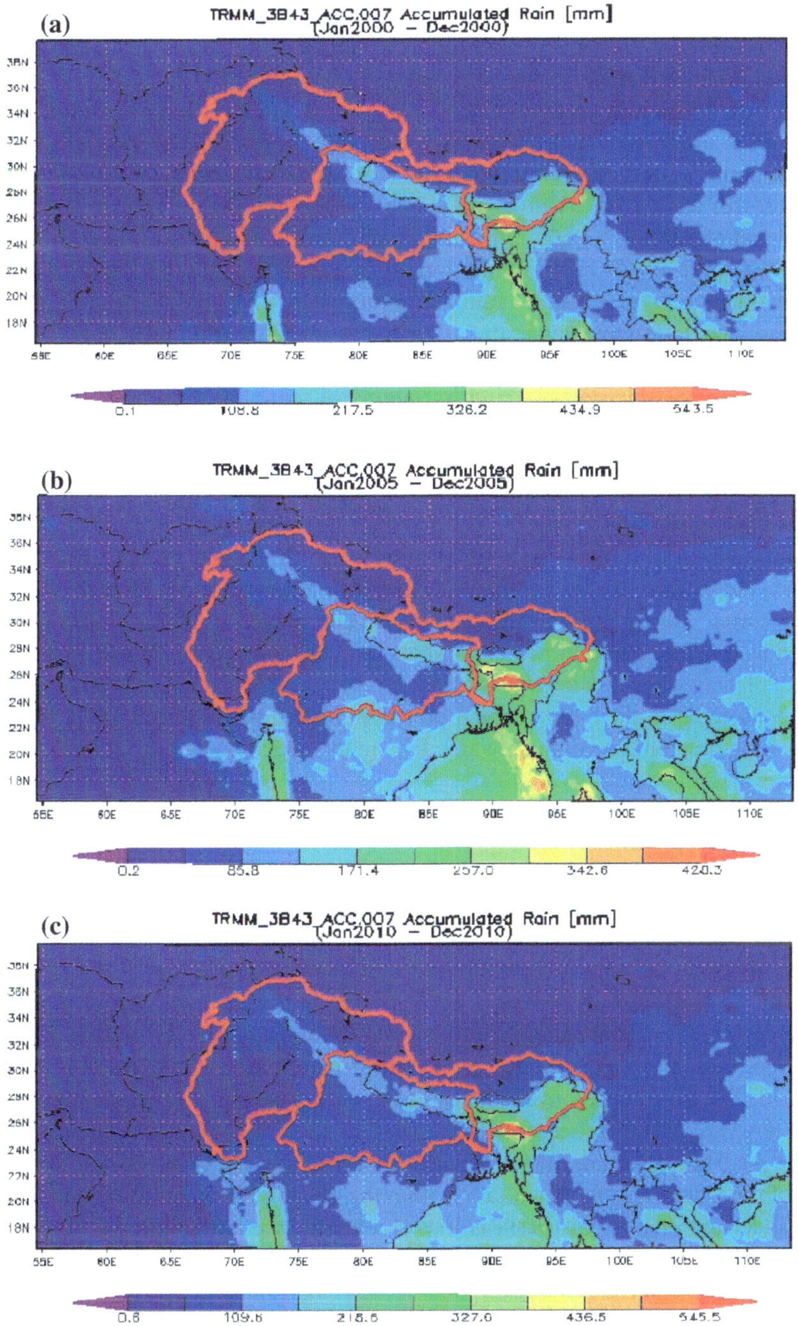

Fig. 4 a–c: Representative annual TRMM rainfall data for the period 2000–2010. The drainage basin boundary of the Ganges, Indus and Brahmaputra are superimposed on TRMM rainfall data

(Fig. 5). A very wide variation across the basin in the annual rainfall is evident. The upper and lower reaches of the Indus River are rain deficient with arid conditions (<300 mm) in the major southern half of the basin.

During the periods of low rainfall (e.g. year 2000; Fig. 5) this region shows drought conditions (annual rainfall <100 mm; Fig. 5). The bulk of the precipitation in this basin is concentrated in a narrow northwest-southeast trending domain in the northern half of the basin where annual precipitation is generally >1,000 mm with local regions of >1,600 mm annual rainfall (Fig. 5). As seen from Fig. 1, a significant part of this high rainfall domain lies in the geographic limits of India. The estimated 11-year annual rainfall for the Indus basin is listed in Table 3. Column 1 contains the annual rainfall for the rectangular grid containing the basin while the second column lists the annual rainfall estimated after subtracting the pixels for the extra-basin area. High variability of rainfall is evident in the TRMM image. A perusal of the estimates of annual rainfall substantiates the visual radar data as the variation between the two is ~20 % at 542 mm for the uncorrected data and 434 mm for the corrected data. Year 2000 represents the lowest rainfall (Fig. 5; 338 mm) and the year 2010 the highest (601 mm; Fig. 5 and Table 3).

There appears to be a crude trend of increase in rainfall during this 11-year period. The estimated average annual rainfall of 434 mm is a close match with one of the earlier estimate of annual precipitation (Immerzeel et al. 2010) though it is significantly lower than the other estimate of 1,051 mm (Jain and Kumar 2012). The explanation for this may possibly lie in the fact that the rainfall measurement data considered in the higher annual precipitation estimate considered the rainfall in the high precipitation domain which falls in the geographic boundaries of India.

2.1.2 Ganges Basin

Unlike the Indus basin, large part of Ganges basin lies in India with a subordinate part in Nepal and a small part in Bangladesh (Fig. 1). The basin area is close to that of the Indus basin at ~1 million km^2 though the reported area varies by ~8 % (Table 4). It initiates from the Gangotri in central Himalayas, initially follows a southwest course and after descending to the Indo-gangetic plains takes a southeast course to the bay of Bengal for the most parts of its ~2500 km long course (Fig. 1). The two recently available estimates of the annual precipitation are comparable at 1,071 mm (Immerzeel et al. 2010) to 1,097 mm (Jain and Kumar 2012) (Table 4). The data from the latter work appeared to have been derived from the part of the Ganges basin which falls within the geographic boundary of India. The total glaciated area of this basin is lowest amongst the three basins at ~1 % and the contribution of the glacial melt water to the total discharge has been estimated to be ~9 % (Table 3). The basin area as derived from SRTM images has been estimated as 951,121,003 km^2 which matches well with that reported by other researchers (Table 4).

The 11-year annual precipitation for northern India and the adjacent regions is shown in the annualized TRMM maps in Fig. 4. The radar precipitation data

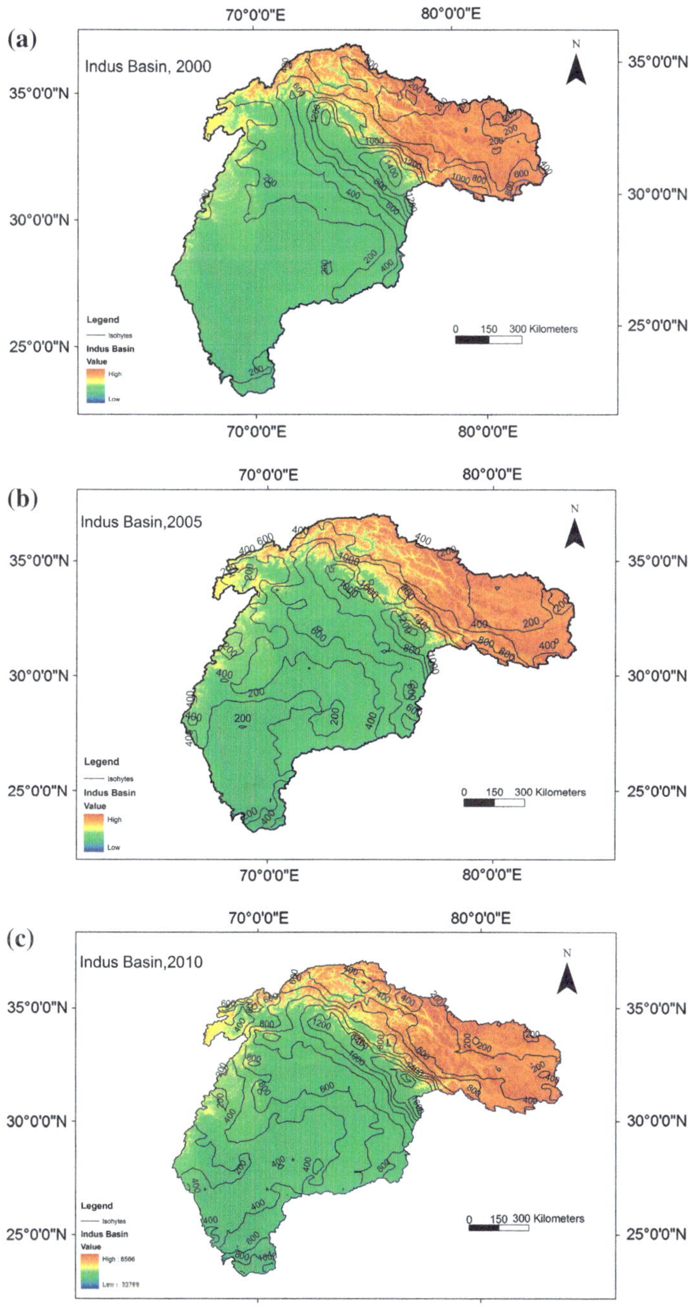

Fig. 5 **a–c**: Representative contoured annual precipitation data for the Indus basin for the years 2000–2010

Table 3 TRMM derived annual precipitation data for the Indus basin

Years	Indus (raw)	Excluded
2000	456.25	338.38
2001	476.01	362.21
2002	400.69	290.87
2003	634.44	512.66
2004	485.97	368.18
2005	575.56	447.81
2006	592.51	482.1
2007	593.7	498.18
2008	598.47	497.98
2009	482.03	374.21
2010	667.81	601.5
Total (mm)	**5,963.44**	**4,774.08**
Average mm/year	**542.13**	**434.01**

shows wide variability in the rainfall data with the northern part of the basin showing significantly high precipitation. Contour diagrams for the annual precipitation estimated from the radar data for three representative years have been plotted on the georeferenced outline of the Ganges basin (Fig. 6). The northern part of the basin is occupied by the Himalayas and has high elevation (Fig. 2). Higher precipitation and very steep precipitation gradient (Fig. 6) indicates strong orographic control. The southwest part of the basin is relatively rain deficient. The estimated 11-year annual rainfall for the Ganges basin is listed in Table 5. Column 1 contains the annual rainfall for the rectangular grid containing the basin while the second column lists the annual rainfall estimated after subtracting the pixels for the extra-basin area.

Compared to the Indus basin, the precipitation in the Ganges basin is dispersed (Fig. 6). Estimates of annual rainfall substantiate the visual depiction of the radar data as the variation between the two is ~0.5 % at 1,034 mm for the uncorrected data and 1,094 mm for the corrected data. Year 2009 represents the lowest rainfall (934 mm) and the year 2003 the highest (1,238 mm; Table 5). There is no secular trend of rainfall variation in this 11-year period. The estimated average annual rainfall of 1,094 mm is a close match with the other estimates (Immerzeel et al. 2010; Jain and Kumar 2012; Table 1).

2.1.3 Brahmaputra Basin

More than half of the Brahmaputra basin, the easternmost of the three basins, lies outside India (Fig. 1). The basin area is nearly half of the other two basins and there appears to be significant variations (~20 %) in the reported basin areas (Table 1). The Brahmaputra River originates in Tibet and flows towards east for a considerable distance, takes a southern turn prior to entering India and thereafter flows westward

Table 4 Compilation of annual average precipitation of the Ganges basin

Total basin area (km^2)	Glaciated area (km^2)	Number of glaciers	Annual basin precipitation in (mm)	Glaciated area in (%)	Glacier melt (%)	Mean discharge (m^3/s)	Estimated ice reserves (km^3)	Total population	References
990,316			1,035	1	3			477,937	Immerzeel et al. (2010)
1,016,124					9.1	18,691		407,466	Jianchu et al. (2009), IHP/HWRP report, IUCN/IWMI et al. (2002), Mi and Xie (2002), Chalise and Khanal (2001), Merz (2004)
861,452			1,051.2						Jain and Kumar (2012)
	9,012	7,963					793.53		Bajracharya and Shrestha (2011), ICIMOD

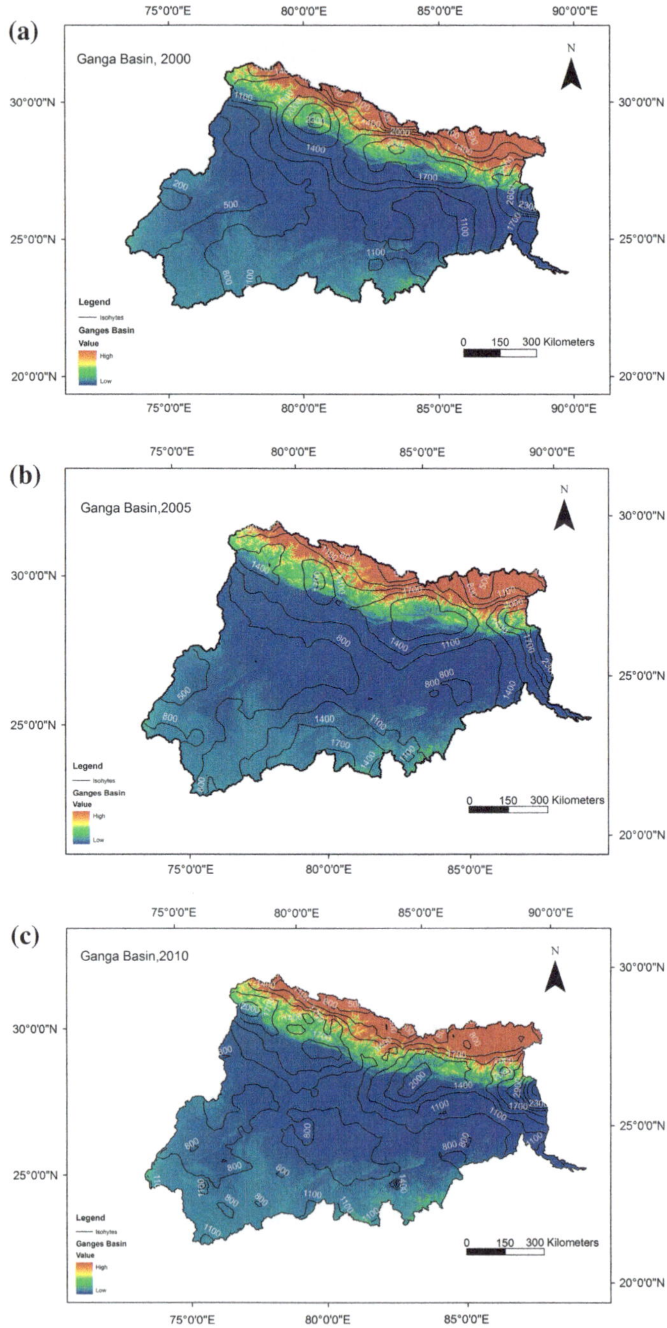

Fig. 6 **a**–**c**: Representative contoured annual precipitation data for the Ganges Basin

Table 5 TRMM derived annual precipitation data for the Ganges basin

Years	Ganges (raw)	Excluded
2000	1,035.5	1,095.5
2001	1,048.8	1,114.5
2002	942.93	980.2
2003	1,146.7	1,238.1
2004	1,034.6	1,077.2
2005	1,037.1	1,089
2006	964.09	1,036.7
2007	1,127	1,181.6
2008	1,120.3	1,181.6
2009	886.92	933.78
2010	1,039.4	1,104.1
Total (mm)	**11,383**	**12,032**
Average mm/year	**1,034.9**	**1,093.9**

till it turns southward to enter Bangladesh and to its merger with the Padma River on its ~2,900 km long channel length (Fig. 1). The two recently available estimates of the annual precipitation are widely variable from 1,071 mm (Immerzeel et al. 2010) to 2,589 mm (Jain and Kumar 2012) (Table 6). The data from the latter work appeared to have been derived from the part of the Brahmaputra basin which falls within the geographic boundary of India. The total glaciated area of this basin is highest amongst the three basins at ~3 % and the contribution of the glacial melt water to the total discharge has been estimated to be ~12.3 % (Table 6). The basin area as recovered from the SRTM images has been estimated as 545,981 km^2 which matches well with that reported by other workers (Table 6).

The 11-year annual precipitation for northern India and the adjacent regions is shown in the annualized TRMM maps in Fig. 4. The radar precipitation data visually shows highest amount of rainfall in this basin amongst the three basins along with widest variability with the northern part of the basin being significantly rain deficient. Contour diagrams for the annual precipitation estimated from the radar data for each of these 11-years have been plotted on the georeferenced outline of the Ganges basin (Fig. 7). The northern part of the basin mainly in the Tibet region is located at high elevation and has high concentration of glaciers (Fig. 2). However, this region has the lowest precipitation (Fig. 7). The eastern and the central part is the region of highest rainfall and it shows variability leading to steep rainfall gradients. This can be attributed to a highly variable orography control as the altitudes in this domain vary significantly (Fig. 2).

The estimated 11-year annual rainfall for the Brahmaputra basin is listed in Table 7. Column 1 contains the annual rainfall for the rectangular grid containing the basin while the second column lists the annual rainfall estimated after subtracting the pixels for the extra-basin area. Compared to the Indus and the Ganges basin, the precipitation in the Brahmaputra basin is significantly higher (Table 7). Estimates of annual rainfall substantiate the visual depiction of the radar data as the variation between the two is ~7 % at 2015 mm for the uncorrected

Table 6 Compilation of annual average precipitation of the Brahmaputra basin

Total basin area (km^2)	Glaciated area (km^2)	Number of glaciers	Annual basin precipitation in (mm)	Glaciated area in (%)	Glacier melt (%)	Mean discharge (m^3/s)	Estimated ice reserves (km^3)	Total population	References
525,797			1,071	3.1				62,421	Immerzeel et al. (2010)
651,335			2,589.2		12.3	19,824		118,543	Jianchu et al. (2009), IHP/HWRP report, IUCN/IWMI et al. (2002), Mi and Xie (2002), Chalise and Khanal (2001), Merz (2004)
194,413									Jain and Kumar (2012)
	14,020	11,497					1,302.63		Bajracharya and Shrestha (2011), ICIMOD

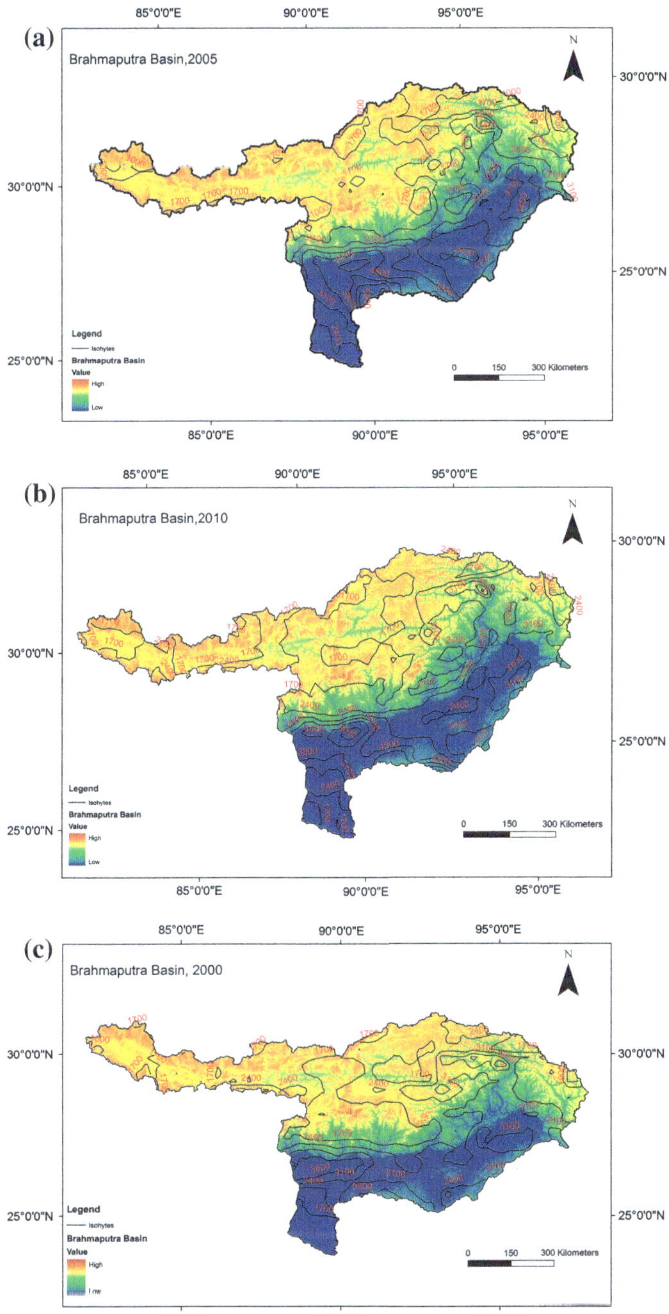

Fig. 7 **a–c**: Representative contoured annual precipitation data for the Brahmaputra basin for the period 2000–2010

Table 7 TRMM derived annual precipitation data for the Brahmaputra basin

Years	Brahmaputra(raw)	Excluded
2000	1,997.9	2,203.7
2001	1,998.2	2,071.8
2002	2,038.4	2,205.4
2003	2,032	2,108.1
2004	2,060.1	2,213.2
2005	1,928.8	2,127.1
2006	1,898.3	1,957
2007	2,139.8	2,243.6
2008	2,053.3	2,199.9
2009	1,783.5	1,863.9
2010	2,231.7	2,377
Total (mm)	**22,162**	**23,571**
Average mm/year	**2,014.7**	**2,142.8**

data and 2,143 mm for the corrected data. Year 2009 represents the lowest rainfall (1,864 mm) and the year 2010 the highest (2,377 mm; Table 7).

There is no secular trend of rainfall variation in this 11-year period. The estimated average annual rainfall of 2,142 mm is widely variable from other estimates (1,071 mm— Immerzeel et al. 2010; 2589 mm—Jain and Kumar 2012; Table 1). It is possible to explain the variation from the higher estimated annual precipitation as Jain and Kumar (2012) did not include the precipitation data of China within which is the lower precipitation is observed leading to a an apparent increase in the estimate. However, the basin wise annual precipitation value for the Brahmaputra basin given by Immerzeel et al. (2010) is unacceptably low as is evidenced in the 11-year TRMM images where this eastern basin shows the highest precipitation (Fig. 4). In the central high rainfall domain in this basin an eastward increase in precipitation is observable (Fig. 7). Higher rainfall is further corroborated by the highest mean discharge in this basin despite it being the smallest (Tables 2, 4 and 6).

3 Results and Discussion

Annual precipitation data for large basins with highly variable topography having precipitation influenced by multiple systems is difficult to assess. Northern India and adjacent terrains are drained by three major basins, the Indus, the Ganges and the Brahmaputra from west to east. More than half of India's population resides within these three basins. Extreme continental topographic variation is seen in these three basins. The precipitation is both solid as well as liquid in all the three basins and there is a significant, though highly variable, contribution of snow and ice melt to the channel water.

The annual precipitation data can be generated using rain gauge measurements if high densities of such measurement stations are available. In the Himalayas sporadically this data is available based on rain-gauge measurements (Shrestha et al. 2000; Singh and Kumar 1995; Jain and Kumar 2012). Storm events related to steep spatiotemporal distribution of precipitation have also been investigated (Barros et al. 2000; Lang and Barros 2002, 2004). However, in mountainous regions of these three basins the rain-gauge measurement data is sparse and the basin scale annual precipitation cannot be reliably derived from this data (Anders et al. 2006; Bookhagen and Burbank 2010). Besides, gradients of precipitation in mountainous regions are often topography controlled but have also been shown to be independent of elevations (Barros and Lang 2003). Thus, soon after availability of radar reflectivity based precipitation data through TRMM several studies on apatiotemporal distribution of precipitation and estimation of annual precipitation came out (e.g. Anders et al. 2006; Barros et al. 2000; Bookhagen and Burbank 2010). However, basin scale rainfall estimation is still debatable for these three basins.

Radar (TRMM) data of 11-years has been used to estimate average annual precipitation for the three basins. Since the estimates are based on the data of the entire basins and cover a statistically significant time of 11-years covering seasonal variations, our estimates are more robust than those available in the literature. The estimated values of 434, 1,094 and 2,143 mm for the Indus, the Ganges and the Brahmaputra basins indicate a general eastward increasing precipitation pattern which is in consonance with the reported measurements of rainfall. As brought out above the highly variable published data is mainly on account of non-consideration of precipitation on basin scale. We can consider three major controls on precipitation in these three basins namely the westerlies, the Indian monsoon and the East Asian monsoon (Webster et al. 1998; Gadgil et al. 2003; Wang 2006). It is debatable whether the heated Tibetan plateau produces a local low pressure cell which forces monsoon to rearrange itself (Bookhagen and Burbank 2010; Gadgil et al. 2003; Webster 1987) or the insulation provided by the orographic barrier of Himalayas to the heated moist air from south to the dry and cold northern latitudes leading to a strong monsoon (Boos and Kuang 2010).

Though the primary aim of this paper is to present the basin scale precipitation data for these three basins, the generated data provides interesting pointers to contribute to this debate. But this will require segregating domains which are primarily controlled by one of the three precipitation influences. The precipitation data of the Indus basin clearly points to the strong influence of the westerlies and lack of influence of the Indian monsoon. It is, therefore, not considered for this purpose. The Ganges and the Brahmaputra basins are contiguous basins (Fig. 1). The northern east-west trending region of high precipitation in the Ganges basin (Fig. 6) continues in the central part of the Brahmaputra basin (Fig. 7) and the intensity of precipitation decreases westward. This regional trend in the direction and intensity of precipitation suggests primary influence of orographic barrier to precipitation wherein the precipitation intensity decreases westward with the decrease in the strength of the summer monsoon. The secondary influence of development of local low pressure cells is reflected in the steep precipitation gradients which do not follow the movement direction of the summer monsoon. Thus, the basin scale precipitation data suggests primary influence of orography and only a secondary influence of high solar insolation.

References

Anders AM, Roe GH, Hallet B, Montgomery DR, Finnegan NJ, Putkonen J (2006) Spatial patterns of precipitation and topography in the Himalaya. In: Willett SD, Hovius N, Brandon MT, Fisher D (eds) Tectonics, climate, and landscape evolution. Geological Society of America Special Paper 398, pp 39–53

Bajracharya SM, Shrestha B (2011) The status of glaciers in the Hindu Kush Himalayan region. ICIMOD, Kathmandu

Barros AP, Lang TJ (2003) Monitoring the monsoon in the Himalayas: observations in the Central Nepal, June 2001. Monthly Weather Review 131(7):1408–1427

Barros AP, Joshi M, Putkonen J, Burbank DW (2000) A study of the 1999 monsoon rainfall in a mountainous region in central Nepal using TRMM products and rain gauge observations. Geophys Res Lett 27(22):3683–3686

Bookhagen B, Thiede RC, Strecker MR (2005) Late quaternary intensified monsoon phases control landscape evolution in the northwest Himalaya. Geology 33(2):149–152. doi:10.1130/G20982.1

Bookhagen B, Burbank DW (2006) Topography, relief, and TRMM derived rainfall variations along the Himalaya. Geophys Res Lett 33:L08405. doi:10.1029/2006GL026037

Bookhagen B, Burbank DW (2010) Towards a complete Himalayan hydrological budget: spatiotemporal distribution of snowmelt and rainfall and their impact on river discharge. J Geophys Res 115:F03019. doi: 10.1029/2009JF001426

Boss RW, Kuang Z (2010) Dominant control of the South Asian monsoon by orographic insulation versus plateau heating. Nature 463(7278):218–222. doi:10.1038/nature08707

Chalise SR, Khanal NR (2001) An introduction to climate, hydrology and landslide hazards in the Hindu Kush-Himalaya region. In: Tianchi L, Chalise SR, Upreti BN (eds). Landslide hazard mitigation in the Hindu Kush-Himalayas. Kathmandu, ICIMOD

Gadgil S (2003) The Indian monsoon and its variability. Annu Rev Earth Planet Sci 31:429–467. doi:10.1146/annurev.earth.31.100901.141251

Immerzeel WW, van Beek LPH, Bierkens MFP (2010) Climate change will affect the asian water towers. Science 328:1382. doi:10.1126/science.1183188

IUCN/IWMI, Ramsar Convention Bureau, WRI (2002) Watersheds of the world. World Resources Institute, Washington

Jain SK, Kumar V (2012) Trend analysis of rainfall and temperature data for India. Curr Sci 102(1):37–49

Jianchu X, Shrestha A, Eriksson M (2009) IHP/HWRP report on climate change and its impacts on glaciers and water resource management in the Himalayan region

Lang TJ, Barros AP (2002) An investigation of the onset of the 1999 and 2000 monsoon in Central Nepal. Monthly Weather Review 130:1299–1316. doi:10.1175/1520-0493

Lang TJ, Barros AP (2004) Winter storms in the Central Himalayas. J Meteorol Soc Jpn 80(3):829–844

Maurya AS, Shah M, Deshpande RD, Bhardwaj RM, Prasad A, Gupta SK (2011) Hydrograph separation and precipitation source identification using stable water isotopes and conductivity: River Ganga at Himalayan foothills. Hydrol Process 25:1530–1531

Merz J (2004) Water balances, floods and sediment transport in the Hindu Kush-Himalayas. University of Bern, Bern, pp 51–62

Mi D, Xie Z (2002) Glacier Inventory of China. Xi'an Cartographic Publishing House, Xi'an

Shrestha AB, Wake CP, Dibb JE, Mayewski PA (2000) Precipitation fluctuations in the Nepal himalaya and its vicinity and relationship with some large scale climatological parameters. Int J Climatol 20:317–327

Singh P, Kumar N (1995) Effect of orography on precipitation in the western Himalayan region. J Hydrol 199:183–206

Webster PJ (1987) The elementary monsoon. Wiley, New York, p 32

Webster PJ, Magana VO, Palmer TN, Shukla J, Tomas RA, Yanai M, Yasunari T (1998) Monsoons: processes, predictability, and the prospects for prediction. J Geophys Res 103(C7):14451–510. doi:10.1029/97JC02719

Climate Change in the Northwestern Himalayas

Mahendra R. Bhutiyani

Abstract The paper examines the magnitude of warming in Northwestern Himalaya during the period from 1866 to 2006. The analyses of the temperature data show an average rate of increase of about 1.1 °C/100 years during this period. Warming effect is particularly significant during the winter season. Winter temperature has shown an elevated rate of increase (1.4 °C/100 years) than the monsoon temperature (0.6 °C/100 years), due to rapid increase in both, the maximum as well as minimum temperatures, with the maximum increasing much more rapidly. Statistically significant decreasing trends (at 95 % confidence level) in the monsoon and overall annual precipitation during the study period are indicated. In contrast, the winter precipitation has shown an increasing but statistically insignificant trend (at 95 % confidence level). Rising winter air temperatures have caused decreasing snowfall component in total winter precipitation on the windward side of the Pirpanjal Range. The studies also indicate reduced the duration of winter by about two weeks in the last three decades. Role of anthropogenic activities influencing climate change in last three decades can not be ruled out.

Keywords Climate change · Warming · Temperature · Precipitation · Northwestern Himalaya

1 Introduction

The Earth's climate has undergone many changes in the geological past. It was the study of these changes in the 19th Century which led to the postulation of the concept of existence of former glacial and inter-glacial periods (Brönnimann 2002). This discovery sparked off a discussion on the reasons behind them and consequently, many scientific studies were initiated to understand the driving

M.R. Bhutiyani (✉)
Defence Terrain Research Laboratory, Metcalfe House, Delhi, India
e-mail: mahendra_bhutiyani@yahoo.co.in

R. Joshi et al. (eds.), *Dynamics of Climate Change and Water Resources of Northwestern Himalaya*, Society of Earth Scientists Series,
DOI 10.1007/978-3-319-13743-8_8

mechanisms. Considerable data have been amassed since then, from more and more lines of evidence—instrumental records, historical records, proxy records such as dendrochronology based on tree-ring analysis, lake, continental and marine sediments, fossils and pollen grains. Attempts have also been made to reconstruct past climatic fluctuations by studying the response of the glaciers and relative concentration of oxygen isotopes from trapped air samples in the ice-cores drilled in areas like Greenland, Tibetan Plateau and the Himalaya.

Studies have confirmed two types of driving mechanisms or forcings that have caused the climate changes; natural (solar and volcanic) and anthropogenic (human-activity related) (IPCC 1995, 2001, 2007). While variation in solar radiation received on the surface of the Earth due to changes in its orbital parameters (on the time-scales of tens and thousands of years) cause solar forcing, episodes of volcanic eruptions injecting large amounts of sulphur dioxide aerosols in the upper air have caused periodic cooling of the Earth's lower atmosphere. Anthropogenic forcing is the climate change brought about by man's activities. Excessive burning of fossil fuels and changes in land use patterns have increased concentration of gases like CO_2, methane, nitrous oxide, etc. in the atmosphere and have caused the greenhouse effect. Because of their natural origin, solar and volcanic forcings are unavoidable. It is the anthropogenic forcing which has been a matter of great concern. It is believed to be responsible for significant climatic changes that have occurred all over the world in the recent past, with alarming consequences for the mankind.

The climatic changes experienced during the periods of solar forcing were gradual, because they were on fairly long time-scales, spread over tens and thousands of years and on some occasions, a few hundred thousand years. Their gradual nature provided adequate time for the flora and fauna to adapt and undergo suitable changes and flourish in new environment. They even led to the extinction of some species. But the rapidity with which these changes have been experienced in the last two and half centuries has raised certain doubts about their genesis, making it difficult to attribute them to natural causes alone. Coincidentally, this period was also marked by an industrial revolution that started in Europe and had its repercussions all over the globe. Rapid industrialization led to an increase in fossil fuel consumption and in the concentration of greenhouse gases in the atmosphere. Consequent rise in global air temperature, by a modest estimate, of about 0.5 °C to 1.1 °C/100 years in the last 150 years (IPCC 2001, 2007), has led to radical changes in the way we receive precipitation and discharge in our rivers. Probable consequences of these changes are likely to manifest in sea-level rise due to enhanced melting of polar ice-caps and mountain glaciers the world over and short-term increase and long-term decrease in discharge in the rivers (Bhutiyani et al. 2008).

2 Climate Change and Himalayan Mountains

The impact of climate change can be best studied in regions which experience stressed climatic conditions such as deserts and high altitude mountainous areas affected by permafrost conditions (e.g. Tibet, Siberia, and Antarctica etc.). Such areas are characterized by arid climate with scanty precipitation and extreme temperature regimes. The evidences of temporal climate variations, reflected in the deposition of sediments, pollen grains, etc. in deserts and changes in the relative concentration of O^{18} and O^{16} isotopes in glacier and permafrost ice are likely to be better preserved in these areas rather than in regions characterized by heavy rainfall and moderate temperature regimes. Mountain areas such as the Himalayas, the Alps, the Andes, the Rockies etc., because of their great altitudinal range within short distances and high biodiversity, amplify the effects and exhibit climatic regimes that are similar to those of widely separated latitudinal belts (Liu and Chen 2000; Thompson et al. 2000). Besides this, the river systems in mountainous areas, being a key element of hydrological cycle, show good impacts of the shifts in climate. They also bring about the disruption of existing socio-economic structures of population inhabiting their basins (Beniston et al. 1997).

The climate in the Himalaya exercises a dominant control over the meteorological and hydrological conditions in the Indo-Gangetic plains as majority of rivers derive significant portion of their discharge from seasonal snowmelt and/or melting of Himalayan glaciers. In view of their overall importance in the context of Indian sub-continent, the study of long-term and short-term climate changes in the Himalaya and their perilous impacts on its fragile eco-system assumes importance. Because of the vast spread and large variations in hydro-meteorological conditions in different parts and impracticability of carrying out such work for the entire Himalaya, this paper covers the northwestern portion only, comprising of the states of Jammu and Kashmir and Himachal Pradesh (Fig. 1). This region, besides being influenced by, more or less, similar meteorological conditions, has a significant concentration of glaciers in its river basins, and better network of meteorological stations as compared to the central and eastern Himalaya.

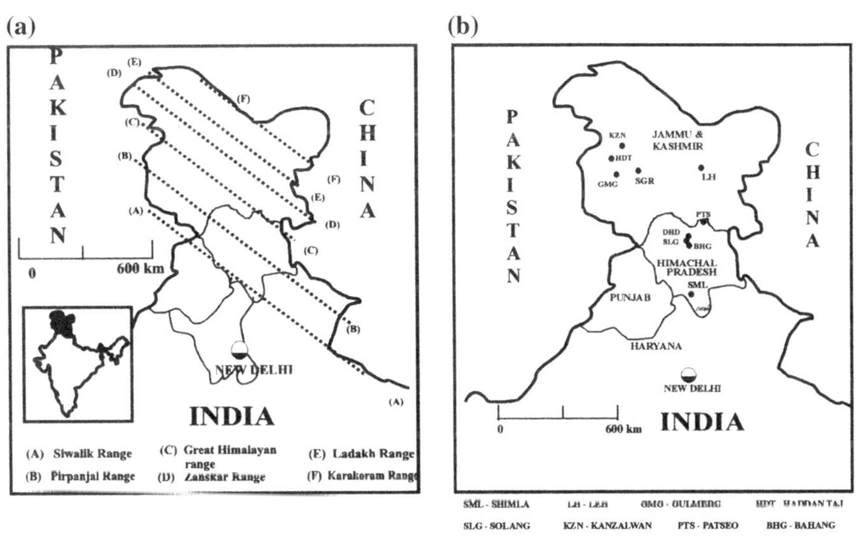

Fig. 1 Map of the northern western Himalaya showing approximate location of various range (**a**) and the meteorological stations (**b**)

3 Magnitude of Climate Change in the NW Himalaya

The climate in northwestern Himalaya (NWH) is influenced by the western disturbances during the winter months from October to May and southwest monsoon from July to September. Precipitation during monsoon period is highest in Siwalik and Pirpanjal Ranges and it reduces as one traverses northwards into the Great Himalaya, Zanskar, Ladakh and Karakoram ranges (Rakhecha et al. 1983). Significant variation is also observed in annual winter snowfall due to western disturbances in the NWH as various ranges in the NWH receive different amounts of snowfall ranging from about 100 to >1,600 cm (snow depth). It is maximum in the Pirpanjal Range and decreases as one goes northwards. Because of the variation in the mean air temperature, the changes occur in the percentage of solid precipitation (snowfall) to rainfall, duration of seasonal snowcover, snow settlement (densification) rates and ablation rates in different ranges of the NWH (Mohan Rao et al. 1987; Bhutiyani 1992).

As compared to some studies on high elevation regions round the globe such as by Diaz and Bradley (1997), Beniston et al. (1997), Beniston (2003), Diaz et al. (2003), Rebetez (2004), very few studies have been carried out on the fluctuations in the climate in the Himalayan Mountains, primarily because of inaccessible terrain and inadequate database. Using the instrumental records, a few studies have examined the rainfall and temperature variations in the Nepal Himalaya and the Tibetan Plateau (Li and Tang 1986; Seko and Takahashi 1991; Borgaonkar et al. 1996) and Upper Indus Basin in Karakoram Himalaya (Fowler and archer 2006). Temperature trends at Katmandu in Nepal Himalaya and the Kosi Basin in Central Himalaya have been studied from the point of view of long-term trends (Sharma et al. 2000). The precipitation trends in the western Himalaya and northwestern Himalaya have also been studied in the last century (Borgaonkar et al. 1996; Bhutiyani et al. 2000, 2004, 2007, 2009; Borgaonkar and Pant 2001; Yadav et al. 2004). Based on proxy data of the ice-cores from Tibetan Plateau and tree-ring analyses from the hill regions of Uttaranchal, some authors have attempted to reconstruct the past climatic conditions during the last few centuries (Pant and Borgaonkar 1984; Liu and Chen 2000; Thompson et al. 2000).

With a view to understand climate change in NW Himalaya on a yearly and decadal basis better, a systematic and detailed study was undertaken to analyze and evaluate climatic as well as hydrological trends in the NWH using instrumental data. The main results and findings of the study are summarized below.

3.1 Temperature Variations

The analyses of the temperature data show that significant increasing trends exist in annual temperature in almost all three main stations namely, Shimla, Srinagar and Leh in the northwestern Himalaya (NWH) in the last century. The

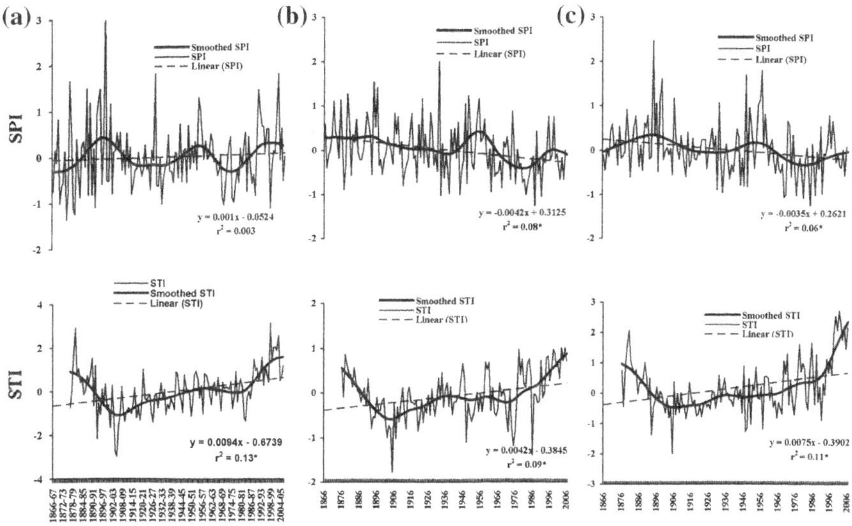

Fig. 2 Temporal variation of winter (**a**), monsoon (**b**) and annual (**c**) standardised precipitation index (SPI) and standardised temperature index (STI) in the northwestern Himalaya (NWH) during the period 1866–2006

annual air temperature has shown an increase of about 1.1 °C during this period. Warming effect is particularly significant during the winter season. For northwestern Himalayan region as a whole, average winter temperature has shown an elevated rate of increase (1.4 °C/100 years) than the monsoon temperature (0.6 °C/100 years) during the period from 1866 to 2006 (Fig. 2). Increase in winter air temperature during three decades is unusually high (about 4.4 °C), as compared to an average rate of about 1.4 °C/100 years in the entire last century. The 'warming' in the NWH has been primarily due to rapid increase in both, the maximum as well as minimum temperatures, with the maximum increasing much more rapidly. Consequently, the diurnal temperature range (DTR) has also shown a significantly increasing trend in both winter and monsoon seasons in the last century (Bhutiyani et al. 2007, 2009). This is in contrast to the findings of studies in the Alps and Rockies (Beniston 1997; Brown et al. 1992) and similar observations on the global scale (Karl et al. 1995), where the minimum temperatures have increased at a higher rate.

Based on the analyses of the temperature data, three different epochs/periods were identified. An episode of comparatively higher (above-average) temperatures from 1876 to 1892 was followed by three more periods of temperature variation. Below-average mean air temperature persisted from 1893 to 1939 indicating a cooler episode, followed by a period of relatively stable/average temperatures till around 1969. The periods from 1969 to 1990 and from 1991 till 2006 are characterized by above-normal temperatures indicating warmer episodes. Temperature seems to have increased at markedly different rates during these two periods. The rate of increase appears to be highest since 1991 as compared to the period prior to 1991

Table 1 Linear trends in monthly air temperatures during winters in the NWH in last three decades

Station	Altitude in m	Data span	Months					
			Nov	Dec	Jan	Feb	Mar	Apr
Bahang	2,192	1977–1978 to 2009–2010	(+)	(+)	(+)*	(+)	(+)*	(+)
Kanzalwan	2,440	1996–1997 to 2009–2010	(+)*	(+)*	(+)	(+)	(+)*	(+)*
Solang	2,480	1996–1997 to 2009–2010	(+)	(−)	(−)*	(−)	(+)	(+)
Gulmerg	2,800	1996–1997 to 2009–2010	(+)*	(−)	(−)	(+)	(+)	(−)
Dhundi	3,050	1989–1990 to 2009–2010	(+)	(+)	(−)	(−)	(+)*	(+)
Haddan Taj	3,080	1996-1997 to 2009-2010	(+)	(+)	(−)	(−)	(+)*	(−)
Patseo	3,800	1996–1997 to 2009–2010	(+)	(+)	(−)	(+)	(−)	(−)

(+) Increasing trend
(−) decreasing trend
* significant at 95 % confidence level

(Bhutiyani et al. 2009), indicating unusual warming in last two decades. This is also confirmed by the analysis of short-term data available for seven stations for the recent decades.

With regard to variation in monthly air temperatures during winter in the last three decades, the studies have indicated non-uniform rate of increase through the winter. Although the beginning of winter (November) has shown an increasing, but statistically insignificant trend, the onset of spring (March) has been marked by substantial warming (Table 1).

3.2 *Precipitation Variations*

Certain variations have occurred in precipitation patterns on the global scale in response to rising temperatures and resultant changes in evaporation from the oceans (Srivastava et al. 1992; Fallot et al. 1997; Zhai et al. 1999). Because of its high temporal and spatial variability and high sensitivity to circulation characteristics, precipitation has rarely been studied in as much details as temperature, as an index of climatic change (Thapliyal and Kulshreshta 1991; Srivastava et al. 1992).

Present study shows a statistically significant decreasing trend (at 95 % confidence level) in the monsoon and overall annual precipitation during the study period. In contrast, the winter precipitation has shown an increasing but statistically insignificant trend (at 95 % confidence level) (Fig. 2). This is generally in good agreement with the results of other studies carried out in western parts of Himalayan foothills (Borgaonkar et al. 1996), Nepal Himalaya (Shtreshta et al. 2000) and in Upper Indus Basin in the Karakoram Himalaya (Archer and Fowler 2004). It can also be seen from the above data that during the period under study, episodes of above-average and below-average winter and monsoon precipitation almost alternated each other with a periodicity varying from 20 to 60 years.

With regard to last few decades, it is seen that whereas monsoon precipitation has remained below-average from 1965 to 2006, winter precipitation has been above-average during the period between 1991 and 2006.

4 Winter Warming and Its Relationship with Winter Snowfall

Although winter precipitation was above-average during the period between 1991 and 2006, studies have shown that rising winter air temperatures have caused decreasing snowfall component in total winter precipitation. This effect is more prominent on the windward side of the Pirpanjal Range and to a lesser extent, some portions on the leeward side. Increasing temperatures during the months of November and March during last three decades probably point towards late onset of winter and early advent of spring season in the NWH. The studies have also indicated that the onset of winter has been delayed by about 2 days per decade and onset of spring has been advanced by about 3 days per decade. This has effectively reduced the duration of winter and consequently the snowfall duration period by 5–6 days per decade and approximately by about 2 weeks in the last three decades (Bhutiyani et al. 2009). Identical results have also been reported from studies in Upper Indus Basin in Karakoram Himalaya (Archer and Fowler 2004), Nagaoka in Japan (Nakamura and Shimizu 1996), the Swiss Alps (Beniston 1997; Laternser and Schneebeli 2003) and Bulgarian mountainous region (Petkova et al. 2004; Brown and Petkova 2007).

Effects of winter warming in last three decades are visible on the Eurasian landmass as a whole. Temporal variation of Eurasian Snowcover Area (ESCA) in March (Data source: Brown 1997, 2002) and winter mean air temperature in the NWH (Fig. 3) demonstrate insignificant trend in the variation of ESCA from 1922 till late-1960s. Consequent decrease in ESCA thereafter is marked by a period of rapidly increasing winter air temperatures in the NWH, indicating a direct inverse relationship between these two parameters. Depleting snowcover area as a result of rising winter temperatures in last three decades may have further amplified the magnitude of winter warming in the Himalayas.

Snow, being a highly reflective material, looses back a large portion of incoming radiation to the atmosphere, thus making a very small portion of energy available for transfer to the ground below. It also acts as a thermal insulator between the ground and the atmosphere, inhibiting the heat transfer between them by conduction and convection.

As more and more land area gets exposed because of decreasing snowcover, larger amount of energy is now available for heating the ground. Consequent higher terrestrial radiation because of elevated ground surface temperatures and higher energy from incoming shortwave radiation increase the net energy balance of the area, which further raises the air temperature of the contiguous areas giving

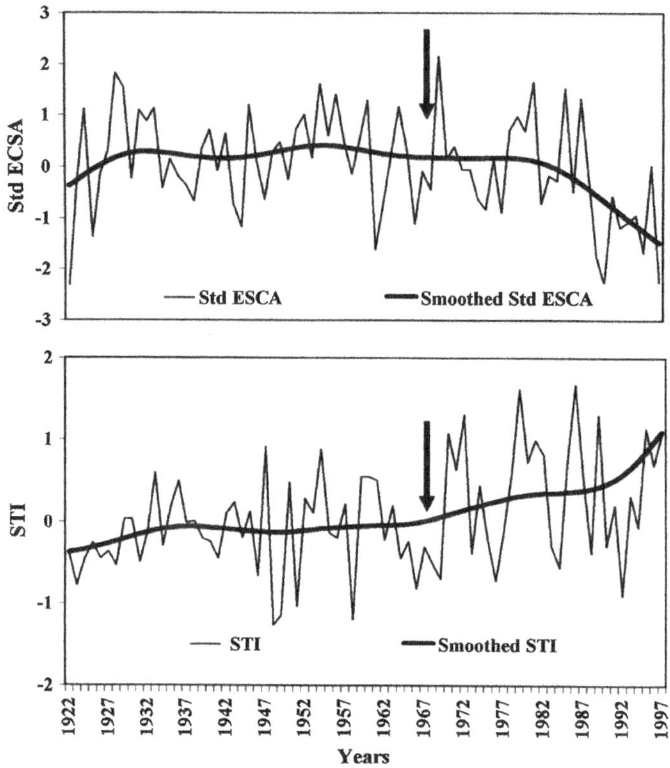

Fig. 3 Temporal variation of standardised Eurasian snowcover Area (ESCA) (March) (*Data Source* Brown (2002)) and winter standardised temperature index (STI) in the northwestern Himalaya (NWH) during the period 1922–1997. An onset of period of rapidly increasing winter air temperature and decreasing Euarsian Snowcover Area (ESCA) is indicated by *black arrows*

rise to a positive 'feedback mechanism'. This effect, which is similar to 'Urban Heat Island' phenomenon generally associated with highly polluted cities, could be termed as 'Mountain Heat Island' effect.

5 Possible Role of Anthropogenic Activities

The studies have demonstrated that although, the temperatures continued to increase from the beginning of the last century, the epochal behaviour of the precipitation ensured presence of comparatively cooler and warmer periods till the early-1970s. The analysis of the diurnal temperature range (DTR) data shows regular periodicity. This periodicity, however, breaks after mid-1970s (Bhutiyani et al. 2007).

It is evident from the foregoing discussion that some natural extra-regional factors such as the Quasi-Biennial Oscillations (QBO) on higher frequency scale of few years and the sunspot activity on the comparatively smaller frequency scale of multi-decades appear to be largely responsible for the precipitation variation in the NWH till early-1970s. The equatorial eastern and central Pacific sea surface temperature (SST) and the ENSO related events had a very limited role to play in these fluctuations. Although, the temperatures continued to increase from the beginning of the last century, the epochal behaviour of the precipitation ensured presence of comparatively cooler and warmer periods (Krishna Kumar et al. 1999). The periods of excess (deficient) annual precipitation, with overall increase (decrease) in cloud-cover, were associated with lower (higher) temperatures, because of the decrease (increase) in net radiation balance. A remarkable feature from the standpoint of the climate change in the NWH is that these tele-connections appear to have weakened considerably in the last three decades i.e. after the early-1970s (Fig. 4) (Krishna Kumar et al. 1999; Baines and Folland 2007; Bhutiyani et al. 2007, 2009).

This convincingly indicates the waning effect of the natural factors in this period. The rise in air temperature has continued unabatedly in this period with both maximum and minimum temperature increasing at an alarming rate. This 'warming' is unusually high and it is difficult to be fully accounted for by the natural forcings alone, as discussed above and there appear to be some additional factors, which may have played a significant role (Easterling et al. 1997; Crowley 2000). One of the external factors could be the increasing concentration of greenhouse gases in the atmosphere. The largest sources of the production of these greenhouse gases are the anthropogenic activities related to rapid industrialization and urbanization. Therefore, monitoring of changes in population and land use patterns and the greenhouse gases emissions may provide an insight into the probable causes of the climatic change in the NWH.

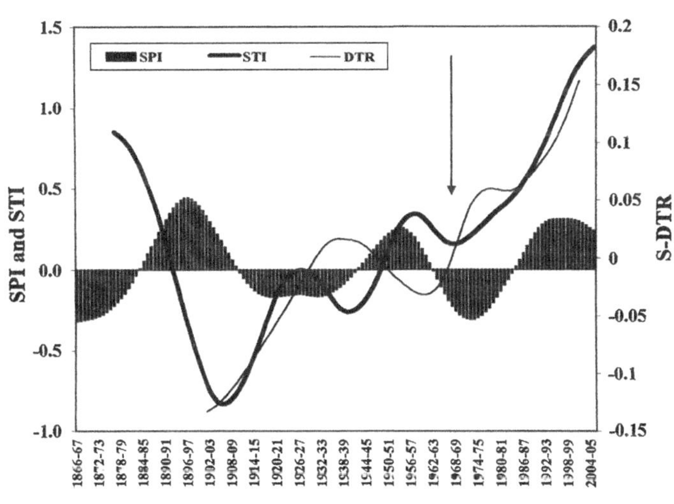

Fig. 4 Relationship between Winer standardised precipitation index (SPI), the standardised temperature index (STI), and standardised diurnal temperature range (S-DTR) in the NWH. *Balck arrow* indicates the period when tele-connections appear to have broken

6 Conclusions

The study has confirmed conclusively that the climate change in northwestern Himalayas is an inevitable reality today and not a myth anymore. The region has 'warmed' significantly during the last century at a rate, which is disturbingly higher than the global average. Unlike other high mountainous regions such as the Alps and Rockies, where the minimum temperatures have increased at a higher rate, the rise in air temperature in the northwestern Himalaya (NWH) has been primarily due to rapid increases in both, the maximum as well as minimum temperatures, with the maximum temperature increasing more rapidly. With regard to precipitation, statistically significant decreasing trends in the monsoon and overall annual precipitation are observed during the study period. In contrast, the winter precipitation has shown an increasing but statistically insignificant trend. Rising winter air temperatures have caused decreasing snowfall component in total winter precipitation, particularly on the windward side of the Pirpanjal Range. The studies have indicated reduction in effective duration of winter by about 2 weeks in the last three decades.

The present study has demonstrated the existence of possible tele-connections between the extra-regional factors such as the Quasi-Biennial Oscillations (QBO), the sunspot activity etc. and the precipitation variation in the NWH till early-1970s in the last century. However, post-1970s, these links appear to have grown weaker considerably, signifying the diminishing effect of the natural forcings during this period and indicating a vital role played by other factors, such as increasing concentration of greenhouse gases in the atmosphere.

References

Archer DR, Fowler HJ (2004) Spatial and temporal variations in precipitation in the Upper Indus Basin, global teleconnections and hydrological implications. Hydrol Earth Syst Sci 8(1):47–61

Baines PG, Folland CK (2007) Evidence for rapid global climate shift across the late 1960s. J Clim 20:2721–2744

Beniston M (1997) Variation of snow depth and duration in the Swiss Alps over the last 50 years: links to changes in large-scale climatic forcings. Clim Change 36:281–300

Beniston M (2003) Climatic change in mountainous regions: a review of possible impacts. Clim Change 59:5–31

Beniston M, Diaz FD, Bradley RS (1997) Climatic change at high elevation sites: an overview. Clim Change 36:233–251

Bhutiyani MR (1992) Avalanche problems in Nubra and Shyok valleys in Karakoram Himalaya, India. J Inst Military Eng India 3:3–5

Bhutiyani MR, Kale VS, Thakur DS, Gupta NK (2004) Variations in winter snowfall precipitation and snow depth patterns in the northwestern Himalaya in the last century: a fallout of global warming? In: Proceedings of international symposium on snow and its manifestations, Manali, India

Bhutiyani MR, Kale, VS, Pawar, NJ (2000) Variations in Glacio-hydrological characteristics of some northwestern Himalayan river basins in this century. In: Proceedings of national seminar on geodynamics and environment management of Himalaya, pp 130–138

Bhutiyani MR, Kale VS, Pawar NJ (2007) Long-term trends in maximum, minimum and mean annual air temperatures across the northwestern Himalaya during the 20th Century. Clim Change 85:159–177

Bhutiyani MR, Kale VS, Pawar NJ (2009) Climate change and the precipitation variations in the northwestern Himalaya: 1866–2006. Int J Climatol. doi:10.1002/joc.1920

Bhutiyani MR, Kale VS, Pawar NJ (2008) Changing streamflow patterns in the rivers of northwestern Himalaya: implications of global warming in the 20th century. Curr Sci 95(5):618–626

Borgaonkar HP, Pant GB (2001) Long-term climate variability over monsoon Asia as revealed by some proxy sources. Mausam 52:9–22

Borgaonkar HP, Pant GB, Rupa Kumar K (1996) Ring-width variations in Cedras deodara and its climatic response over the western Himalaya. Int J Climatol 16:1409–1422

Brönnimann S (2002) Picturing climate change. Clim Res 22:87–95

Brown RD (1997) Historical variability in North Hemisphere spring snow covered area. Ann Glaciol 25:340–346

Brown RD (2002) Reconstructed North American, Eurasian, and Northern Hemisphere snow cover extent, 1915–1997. National Snow and Ice Data Center. Digital media, Boulder

Brown RD, Petkova N (2007) Snowcover variability in Bulgarian mountainous regions. Int J Climatol 27:1215–1229

Brown TB, Barry RG, Doesken NJ (1992) An exploratory study of temperature trends for Colorado Paired Mountain-High plains Stations. In: American Meteorological Society sixth conference on mountain meteorology, Portland, pp 181–184

Crowley TJ (2000) Causes of climate change over the past 1,000 years. Science 289:270–276

Diaz HF, Bradley RS (1997) Temperature variations during the last century at high elevation sites. Clim Change 36:253–279

Diaz HF, Grosjean M, Graumlich L (2003) Climate variability and change in high elevation regions: past, present and future. Clim Change 59:1–4

Easterling DR, Horton B, Jones PD, Peterson TC, Karl TR, Parker DE, Salinger JM, Razuvzyev V, Plummer N, Jamason P, Folland CK (1997) Maximum and minimum temperature trends for the globe. Science 227:364–365

Fallot JM, Barry RG, Hoogstrate D (1997) Variation of mean cold season temperature, precipitation and snow depths during the last 100 years in the former former Soviet Union (FSU). Hydrol Sci 42:301–327

Fowler HJ, Archer DR (2006) Conflicting signals of climatic change in the Upper Indus Basin. J Clim 19:4276–4293

IPCC (1995) The science of climate change. Cambridge University Press, UK

IPCC (2001) Climate Change. The IPCC third assessment report, vols I (The scientific basis), II (Impacts, adaptation and vulnerability) and III (Mitigation). Cambridge University Press, Cambridge

IPCC (2007) Climate Change 2007: climate change impacts, adaptation and vulnerability, Working Group II Contribution to the Intergovernmental Panel on Climate Change—fourth assessment report summary for policymakers

Karl TR, Knight RW, Plummer N (1995) Trends in high frequency climate variability in the twentieth century. Nature 377:217–220

Krishna Kumar K, Rajgopalan B, Cane MK (1999) On the weakening relationship between the Indian Monsoon and ENSO. Science 284:2156–2159

Laternser M, Schneebeli M (2003) Long-term snow climate trends of the Swiss Alps (1931–1999). Int J Climatol 23:733–750

Li C, Tang M (1986) Changes of air temperature of Qunghai—Xizang plateau and its neighborhood in the past 30 years. Plateau Meteorology 5:322–341

Liu X, Chen H (2000) Climatic warming in the Tibetan Plateau during recent decades. Int J Climatol 20:1729–1742

Mohan Rao N, Rangachary N, Kumar V, Verdhan A (1987) Some aspects of snow cover development and avalanche formation in Indian Himalaya. In: Proceedings of Davos symposium on avalanche formation, movement and effects. International Association of Hydrological Sciences, publication No 162, pp 453–462

Nakamura T, Shimizu M (1996) Variation of snow, winter precipitation and winter air temperature during the last century at Nagaoka, Japan. J Glaciol 42:136–140

Pant GB, Borgaonkar HP (1984) Climate of the hill regions of Uttar Pradesh. Himalayan Research and Development 3:13–20

Petkova N, Koleva E, Alexandrov V (2004) Snowcover variability and change in mountainous regions of Bulgaria, 1931–2000. Meteorol Z 13:19–23

Rakhlecha PR, Kulkarni AK, Mandal BN, Dhar ON (1983) Winter and spring precipitation over the northwestern Himalaya. In: Proceedings of the first national symposium on seasonal snow cover, New Delhi, vol 2, pp 175–181

Rebetez M (2004) Summer maximum and minimum daily temperature over a 3300 m altitudinal range in the Alps. Clim Res 27:45–50

Seko K, Takahashi S (1991) Characteristics of winter precipitation and its effects on glaciers in Nepal Himalaya. Bull Glacier Res 9:9–16

Sharma KP, Moore B, Vorosmarty CJ (2000) Anthropogenic, climatic and hydrologic trends in the Kosi Basin, Himalaya. Clim Change 47:141–165

Shreshtha AB, Wake CP, Dibb JE, Mayewski PA (2000) Precipitation fluctuations in the Nepal Himalaya and its vicinity and relationship with some large-scale climatological parameters. Int J Climatol 20:317–327

Srivastava HN, Dewan BN, Dikshit SK, Prakasha Rao GS, Singh SS, Rao KR (1992) Decadal trends in climate over India. Mausam 43:7–20

Thapliyal V, Kulshrestha SM (1991) Decadal changes and trends over India. Mausam 42:333–338

Thompson LG, Yao T, Mosley-Thompson E, Davis ME, Henderson KA, Lin PN (2000) A high-resolution millennial record of the South Asian monsoon from Himalayan ice cores. Science 289:16–19

Yadav RR, Park WK, Singh J, Dubey B (2004) Do the western Himalaya defy global warming? Geophys Res Lett 31:L17201

Zhai P, Sun A, Ren F, Xiaonin L, Gao B, Zhang Q (1999) Changes in climate extreme in China. Clim Change 42:203–218

Aerosols and Temperature Rise in the Northwestern Himalaya, India

Jagdish Chandra Kuniyal

Abstract Aerosols, the important constituents of our atmosphere, indicate a colloidal system of particulate, gaseous and volatile organic compounds. Aerosols play a significant role in affecting adversely the radiative balance of the Earth as well as the air temperature. Moreover, these not only influence the visibility and overall air quality, but also adversely affect health of living organisms in an ecosystem. In this context, the present attempt at Mohal (1,154 m, 77.12°E, 31.91°N) in the Kullu valley of Himachal Pradesh in the northwestern Himalayan region explains the ever increasing columnar aerosols, their relationship with black carbon (BC) aerosols, impact of local meteorological conditions, long range transport sources and their collective impact on radiative forcing and resultant temperature rise. The aerosol optical depth (AOD) having been under observation for the last half a decade (2006–2010) shows higher values at shorter wavelengths and lower at longer wavelengths. At a representative wavelength of 500 nm, AOD is found to be increasing at the rate of 1.9 % per annum from 2006 to 2010. Overall, AOD values in all the wavelengths (380–1,025 nm) were found between 0.238–0.242, reflecting an increasing trend at the rate of 0.84 % per annum. The monthly mean concentration of BC aerosols is noticed maximum with 6,617 ng m^{-3} in January, 2010. The pollution loads in terms of AOD values translate into a temperature rise by ~0.54 K day^{-1}. The local as well as transported aerosols together contribute to the existing aerosols in the present study region. The local sources possibly belong to anthropogenic aerosols including vehicular emissions, biomass burning (like fuel wood for cooking), forest fires, open waste burning, etc. While the transported aerosols most probably include fine mineral dust from the desert regions and the sulphate aerosol from the oceanic regions with the movement of air masses prior to the western disturbances and monsoonal winds in the region.

J.C. Kuniyal (✉)
G. B. Pant Institute of Himalayan Environment and Development,
Himachal Unit, Mohal-Kullu, Himachal Pradesh, India
e-mail: jckuniyal@gmail.com

R. Joshi et al. (eds.), *Dynamics of Climate Change and Water Resources of Northwestern Himalaya*, Society of Earth Scientists Series,
DOI 10.1007/978-3-319-13743-8_9

Keywords Aerosol · Aerosol optical depth · Black carbon · Pollution sources · Temperature rise · Northwestern Himalaya

1 Introduction

The study of aerosols, in modern science, is not only interesting but is also growing popular day by day. Aerosols are the colloidal system of particulate, gaseous and liquid pollutants which remain in suspension in the Earth's atmosphere (Iqbal 1983). The most important optical properties of the aerosols is the aerosol optical depth (AOD) which is directly proportional to the magnitude of the attenuation of direct solar radiation by scattering and absorption process (Charlson et al. 1992). If the aerosols remain with higher concentrations in the atmosphere, these may have impacts either on human health or plant life (Liu et al. 1991) or climate impacts causing radiative forcing, temperature rise and ultimately climate change. Fine size particles are primarily generated through anthropogenic activities. The process of gas to particle conversion is responsible for a variety of the other forms of pollution like haze, mist and fog. These pollutions are responsible for respiratory problems in human beings which may become a cause of carcinogen.

The Himalayan region is considered as a white box in terms of availability of data on aerosols which may have long-term effects in temperature rise. It is, therefore, very important to conduct study on aerosols in the sub-mountain region of the Indian Himalaya similar to the Indo-Gangetic Plain (Singh et al. 2008), northwestern Himalaya (Kuniyal et al. 2007, 2009), and other parts of the western, eastern, and coastal regions of the country (Beegum et al. 2008). Himalaya being considered as the third pole after Antarctica and Arctic; studies beyond Indian Himalayan Region (IHR), around the Tibetan plateau (Liu et al. 2008) and other parts of the Hindu-kush Himalayan region are rare. The solar energy reaching at top of the atmosphere (TOA) arrives at the Earth's surface with attenuation mainly due to aerosols and other atmospheric factors. Some amount of the radiant energy is reflected back into the space, while some is absorbed by the aerosols within the atmosphere. However, the reflective and absorptive solar energy depends much on their nature and quality of aerosols in the atmosphere. Sulphate, the oceanic aerosol, has reflective quality (Moorthy et al. 1998; Satheesh et al. 2001), while black carbon, the carbonaceous aerosol produced mostly from the local biomass burning sources, has a heat absorbing quality. So depending on the nature of aerosols, cooling effect mostly near the surface as well as at top of the atmosphere and warming effect in the atmosphere together result in the radiative imbalance. This phenomenon from local to regional level later has implications in existing climate of a particular geographic entity. The process of scattering and absorption of solar irradiance or incoming solar radiation alter the radiation budget of the Earth's atmosphere (Fu and Liou 1992, 1993) which in due course results for changes in climate (IPCC 2007).

Keeping in view of these issues pertaining to aerosols, temperature rise and climate change, the present study is carried out with the objectives to: (i) obtain aerosol optical depth (AOD) in ultra-violet, visible and near infrared wavelengths (380–1,025 nm), (ii) estimate the concentration of black carbon aerosols, (iii) identify the possible local and external pollution sources estimating vehicular influx, etc. and HYSPLIT model of back trajectory analysis respectively, and (iv) monitor aerosol impact on temperature rise and hence on climate change in the region.

2 Experimental Site and Methodology

The observations for AOD, black carbon (BC) and meteorological parameters were carried out at Mohal (1,154 amsl, 77.12°E, 31.91°N), located at 5 km south to Kullu town, in the northwestern Indian Himalayan region. The experimental site is located within the campus of Himachal Unit of G. B. Pant Institute of Himalayan Environment and Development, at Mohal-Kullu, Himachal Pradesh. The distance of the sampling site from the National Highway (NH)-21 is about 288 m, while direction of the experimental site from the NH-21 is in the north–west.

The measurements of AOD, BC and meteorological parameters were carried out using the Multi-wavelength Radiometer (MWR), Aethalometer, and Automatic Weather Station (AWS), respectively. The MWR instrument works on the principle of filter wheel radiometer as described by Shaw et al. (1973). The MWR, developed at Space Physics Laboratory, Vikram Sarabhai Space Centre, Thiruvanthapuram, has ten wavelength bands, centered at 380, 400, 450, 500, 600, 650, 750, 850, 935 and 1,025 nm (Moorthy et al. 1999; Bhuyan et al. 2005; Gogoi et al. 2008, 2009), whereas the field of view of the MWR instrument is 2° (Moorthy et al. 1999). From each set of MWR measurements, the columnar optical depths (τ_λ) were estimated using the Langley technique by making a linear regression fit to Bouger-Lambert-Beer Law (Moorthy et al. 1999) as follows,

$$\ln E_\lambda = \ln E_{o\lambda} + 2 \ln (d_o/d) - m\tau_\lambda \tag{1}$$

where E_λ represents the ground reaching solar radiation, $E_{o\lambda}$ indicates the solar radiation incident at the top of the atmosphere, d_o and d are the mean and instantaneous values of the Sun-Earth distance, m indicates the relative air-mass as a function of Solar Zenith Angle $\leq 70°$.

The τ_λ obtained from Eq. 1 explains the total columnar optical depth as an outcome of different atmospheric extinction processes, i.e., scattering due to air molecules ($\tau_{r\lambda}$), aerosols ($\tau_{p\lambda}$), gaseous (i.e. ozone and nitrogen dioxide) and water vapour absorption ($\tau_{\alpha\lambda}$) (Moorthy et al. 1991, 1996),

$$\tau_\lambda = \tau_{r\lambda} + \tau_{\alpha\lambda} + \tau_{p\lambda} \tag{2}$$

Thereafter, $\tau_{p\lambda}$ is estimated for each observation day using Eq. (2) (Moorthy et al. 1996) as under,

$$\tau_{p\lambda} = \tau_\lambda - (\tau_{r\lambda} + \tau_{\alpha\lambda}) \tag{3}$$

The spectral dependence of AOD is used to compute Ångström exponent (α) and turbidity coefficient (β) in the wavelength interval 0.38–1.025 µm by making a linear regression fit to Ångström Power Law (Ångström 1961),

$$\tau_{p\lambda} = \beta\,\lambda^{-\alpha} \quad (4)$$

The continuous and near-real-time measurements of the mass concentration of black carbon aerosols were carried out using Aethalometer (Model AE-31-ER, Magee Scientific, USA). Aethalometer works based on the attenuation of a beam of light transmitted through a quartz fiber filter tape which continuously collects aerosol sample from ambient air (Raju et al. 2011). The instrument is operated at an average time of 5 min, round the clock with a flow rate of 2.9 L m^{-1}. The BC mass concentration is estimated by measuring the change in the transmittance of a quartz filter tape (Hansen et al. 1984). The optical attenuation (ATN) is explained as follows (Hansen 2005),

$$\mathrm{ATN} = 100 \times \ln(\mathrm{I_o}/\mathrm{I}) \quad (5)$$

where I_o = intensity of the light transmitted through the original filter/blank portion, I = intensity of the light transmitted through the portion of the filter on which BC is collected, 100 corresponds to an aerosol spot that is quite dark grey.

Shortwave (280–4,000 nm) surface solar flux measurements were carried out using Pyranometer (make Kipp & Zonen). The meteorological parameters like temperature, wind speed, wind direction and relative humidity were obtained using the AWS installed at GBPIHED, Himachal Unit, Mohal, Kullu (H.P.). The influence of aerosols in changing shortwave solar radiation were investigated taking into account long range transport using Hybrid Single Particle Lagrangian Integrated Trajectory (HYSPLIT) model (website: http://www.arl.noaa.gov/ready/hysplit4.html) along with Cloud-Aerosol Lidar and Infrared Pathfinder Satellite Observations (CALIPSO) (Winker et al. 2009).

In the present study, the most important input parameters for estimating the aerosol radiative forcing such as AOD, single scattering albedo (SSA) and asymmetry parameter (g) are retrieved from Optical Properties of Aerosol and Clouds (OPAC) database by adopting the external mixing approach (Hess et al. 1998). The OPAC model is run to estimate aerosol optical properties between 0.25 and 4.0 µm. As the measurements of aerosol chemical compositions were not available; we have used a hybrid approach to estimate aerosol optical properties. The zero-order approximation is performed by adopting polluted continental aerosol types (post-monsoon and winter months) and desert aerosol types (pre-monsoon and monsoon months) in the OPAC database. The earlier studies (Guleria et al. 2011a, b) have reported that transport of dust aerosols from the western desert contributes significantly to the existing concentration of coarse mode aerosols during pre-monsoon and monsoon seasons. Besides local sources, anthropogenic aerosols from the polluted continental region of India in post-monsoon and winter also contribute to the large concentration of fine mode aerosols. In view of possible sources of aerosols over Mohal, the choice of polluted continental and desert aerosol types in the OPAC database have been adopted. The OPAC model is run to reconstruct the AOD spectra taking AOD, α, and BC aerosol number concentration as the anchoring points.

The OPAC estimated optical properties of aerosols namely AOD, SSA and *g* are further used to calculate clear-sky radiative forcing in the shortwave region ranging from 0.25 to 4.0 μm by using radiative transfer model (RTM) based on the Discrete Ordinate Radiative Transfer (DISORT) module (Ricchiazzi et al. 1998). The DISORT method was first proposed by Chandrasekhar (1950) and later modified and improved by Liou (1973) and Stamnes et al. (1988). This radiative transfer code uses numerically stable algorithm to solve the equations of plane-parallel radiative transfer in vertically inhomogeneous atmosphere (Ricchiazzi et al. 1998). Following successful applications in modeling the radiative transfer, the DISORT method has become most popular among atmospheric and space research community (Ricchiazzi et al. 1998; Moorthy et al. 2005; Jin et al. 2006; Pant et al. 2006; Babu et al. 2007; Pathak et al. 2010; Singh et al. 2010). The RTM is run to compute plane-parallel radiative transfer in cloud-free days within the atmosphere-land system. The clear-sky shortwave aerosol radiative forcing calculations are performed separately with and without aerosols at 5° zenith interval and are used to determine 24 h averages.

3 Results and Discussion

3.1 AOD—Forenoon and Afternoon

The full day mean AOD values at ten wavelengths (380, 400, 450, 500, 600, 650, 750, 850, 935, 1,025 nm) for the clear-sky days for the last 5 years from 2006 to 2010 were obtained. Based on the study, it was found that AOD values were noticed maximum for the year 2010 and minimum for the year 2007 (Fig. 1a). On average, the AOD value at 500 nm was 0.22 in 2007 which increased up to 0.27 in 2010. This increase was 22 % between 2007 and 2010. The fine size aerosols as AOD within accumulation and fine mode (380–500 nm) are supposed to

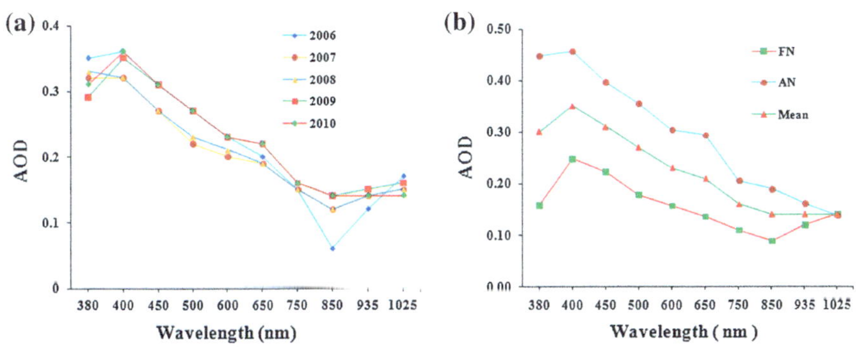

Fig. 1 Columnar aerosols: **a** AOD values (2006–2010), and **b** FN, AN, mean AOD for the year 2010

be produced due to anthropogenic interferences, which ranged from 0.28 to 0.31 AOD, an increase by 10.71 % at the rate of 3.57 % per annum. While the AOD values, on average, for coarse size particles (600–1,025 nm) were found in a range from 0.16 to 0.17, an increase by 6.25 % at the rate of 3.1 % per annum from 2007 to 2010. Primary reasons of increase from 2006 to 2010 are possibly the deforestation, sprawling urban towns (or urbanization), industrialization (like cement plants, etc.), upcoming hydropower projects, forest fires, vehicular emissions, open waste burning, etc. It is found that the AOD values are higher at shorter wavelengths and lower at larger wavelengths indicating dominance of anthropogenic aerosols over natural ones. As a day advances from forenoon to afternoon (FN to AN), temperature of the Earth's atmosphere also increases. As a result, the process of lifting air pollutants or aerosols above the ground due to convective process gets dispersed at higher levels (Kuniyal et al. 2009).

This convective process results in relatively higher concentration of aerosols during afternoon (Fig. 1b). In addition, it is observed that the long range transport of aerosols also increase the atmospheric turbidity level. The cool air mass flow laden with finer size air pollutants at lower height reaches at its peak during afternoon when local air gets warm up due to relatively stronger sunshine and outside cool air takes its place. The highest annual forenoon AOD during the clear sky days at 500 nm was 0.59 on 30 May 2006, while its lowest value was 0.33 on 24 November 2010. At the same time, the maximum afternoon AOD for the clear sky days at 500 nm was 0.76 on 19 March 2010 and lowest value was 0.48 on 22 April 2007 as well as on 31 January 2008. It is therefore observed that AOD values at forenoon are lower than those of observed values at afternoon.

3.2 Variations in Ångström Exponent (α) and Turbidity Coefficient (β)

Ångström parameters such as wavelength exponent (α) and turbidity coefficient (β) indicate size distribution of the scattering particles and total aerosol content respectively. In a majority of the cases, it is found that α and β were inversely proportionate to each other (Fig. 2a). The monthly mean value of α and β for the clear sky days recently in 2010 was also obtained. When α values remain very small or say of the order of air molecules, it may approach up to 4 while in case of very large size particles, it may be as low as 0 (Holben et al. 2001; Pinker et al. 2001). Ångström exponent 'α' as the maximum was 1.54 in the month of August and minimum was 0.54 in the month of March indicating dominance of finer size aerosols in August over the other months of a year. While the maximum value of the 'β' was 0.31 in the month of July and the minimum was 0.07 in the month of August indicating the maximum visibility in the atmosphere in August due to very low turbidity.

One of the reasons for high visibility and low turbidity in the month of August in 2010 may be due to washout effects of rainfall when maximum sixteen rainy days during this month were recorded. Due to increase in α, the fine particle

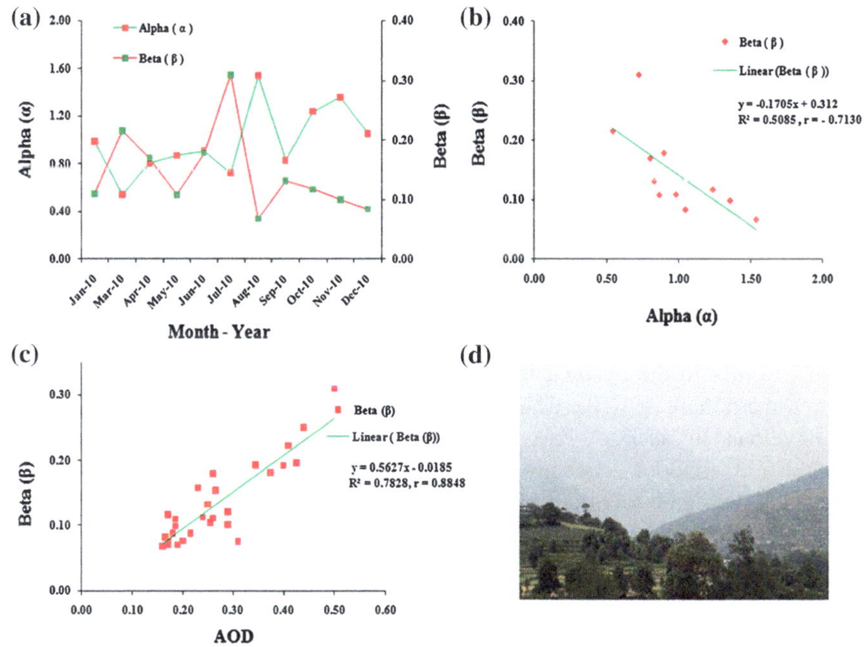

Fig. 2 Ångström parameters: **a** monthly mean values of α and β for the year 2010, **b** correlation coefficient between α and β, **c** correlation coefficient between AOD, β, and **d** a view of hazy atmosphere in the Kullu valley

concentrations increase (Dumka et al. 2008). The peculiar feature in case of fine particles is that these have maximum heat absorbing capacity that result in warming effect for longer duration in the atmosphere. However, increase in β indicates increase in the total particles leading to turbidity in the atmosphere that cause low visibility. While the correlation coefficient between α and β for the clear sky days of the year 2010 unfolds it to be $r = -0.713$ indicating strong negative correlation between these two Ångström parameters. In other words, when α increases, β decreases and vice versa (Fig. 2b).

The correlation coefficient between AOD at 500 nm and β for the clear sky days in 2010 estimated to be $r = 0.88$ indicating strong positive correlation between AOD and β (Fig. 2c). This means that when AOD value increases, β value also increases. In other words, when coarse size particles increase, there is also an increase in AOD values resulting in turbidity conditions in the atmosphere (Fig. 2d).

3.3 AOD and Black Carbon Aerosol

The diurnal variation in BC aerosol mass concentrations is influenced partly by the diurnal vertical increase in the atmospheric boundary layer (ABL) (Ganguly et al.

2006; Pathak et al. 2010) and partly by increasing human activities in the forms of vehicular emissions and biomass burning. The ABL is low during morning time, thereafter gradually rises with increasing sunshine hours up to its maximum level by noon. When sun sets, BC starts to decrease by the evening hours. In early morning hours, the sharp increase in the production of BC aerosol mass concentrations is mainly attributed to the dynamics of the local ABL. Besides, routine activities of the households and commuters in the form of biofuel or biomass/fuel wood burning might also be contributing to its existing level of concentration. After sunrise, the ABL starts to uplift which results in thermal strengthening that lifts up the BC aerosols and eventually breaks the nighttime inversion. This gradual formation of a surface based inversion opposes vertical mixing in the atmosphere causing aerosols to mix in the residual layer with those aerosols near the surface which lead to a sharp increase in the near-surface concentrations.

This effect in such a way caused is known as fumigation (Stull 1998; Fochesatto et al. 2001). Because of this effect, the primary peak generally is observed around 8.00 to 9.00 h Indian Standard Time (IST). As the day advances, the ABL is lifted at its highest level due to warming of the Earth's surface through solar energy and thereafter it gets weakened as the sun sets resulting in low BC production in the ambient air near the surface. This atmospheric process results in sharp decrease in the BC aerosol mass concentrations. During sunset hours, the burden of anthropogenic activities were realized in the form of vehicular influx, open waste burning, burning of crop residues, forest fire, and fuel wood burning to keep the house temperature warm. During winter, it again starts to increase from around 16.00 h IST and reaches at its peak by around 20.00 h IST. Moreover, the radiative cooling of the ground surface results in suppression of turbulent mixing and consequently increases the possibility of aerosol concentrations during early night.

The gradual increase in BC aerosol mass concentrations from around 16.00 h IST is considered to be due to alike anthropogenic activities and gradual formation of a surface based inversion opposing vertical mixing in the atmosphere (Ganguly et al. 2006; Nair et al. 2007). After reaching at its secondary maxima, the BC aerosol mass concentrations decreases gradually because of low anthropogenic activities during this period.

The hourly highest ever BC aerosol mass concentration was observed as 32,447 ng m^{-3} on December 28, 2009 at 8.00 h IST and 33,764 ng m^{-3} on January 12 at 7.00 h IST. The monthly mean values of BC aerosols at Mohal in 2010 are shown in Fig. 3. Its monthly mean concentration was noticed maximum with 6,617 ng m^{-3} in January, 2010 and minimum with 3,259 ng m^{-3} in June, 2010. In general, the BC aerosol mass concentration increases with an increase in human activities like vehicular emissions, biomass burning (fuel wood for cooking), forest fires, etc. The average of FN and AN AOD at 500 nm for a full day was compared with the average of BC aerosol mass concentration over a period of 2010 year for the clear sky days (see Fig. 3). The BC aerosols are primarily from the anthropogenic sources (Badrinath and Kharol 2008), while the AOD sources are from both the sources; natural and anthropogenic.

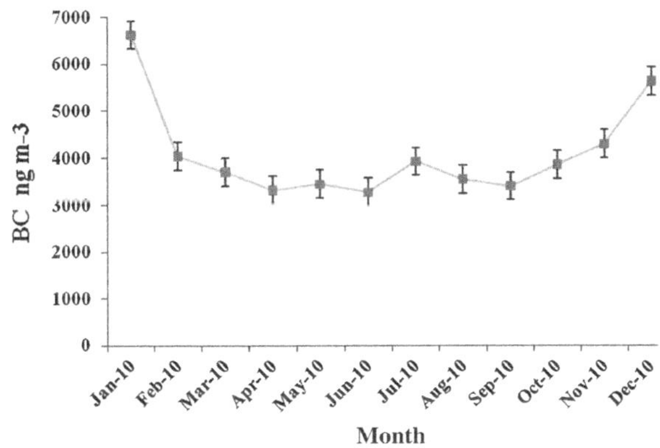

Fig. 3 Black carbon (BC) aerosols concentrations at Mohal-Kullu, 2010

3.4 AOD and Meteorological Parameters

3.4.1 AOD and Temperature

The AOD values obtained at 500 nm were compared with the average values of the temperature for a period of MWR operation during the clear sky conditions in 2010. It is found that the correlation coefficient between these two parameters was weak positive ($r = 0.361$). However, increase in temperature leads to the convective process and aerosols are lifted up from the surface within ultra violet, visible, and infrared spectrums.

3.4.2 AOD, Wind Speed and Wind Direction

The average values of AOD at 500 nm when was compared with the average values of wind speed for the clear sky days in 2010, it was found that the windblown aerosols from the polluted regions reach to our study region. As a result, existing concentration of aerosols also increase in our region. As the wind speed increases, the AOD values also shoot up. The mean values of wind direction were from the south-east and south direction to the present study region.

3.4.3 AOD, Back Trajectory and Satellite Data

For examining the possibility about the transport of trans-boundary pollutants from the external sources to our study region, 7 day's back trajectories were drawn using HYSPLIT model (Draxler and Rolph 2010). The back trajectories were drawn at the three different heights 3,500 m, 5,500 m and 7,500 m above the ground level (Fig. 4a). The trajectories drawn at 3,500 m AGL is influenced by the activities generating pollutants within the valley, whereas trajectories at 5,500 and 7,500 m AGL are taken into account considering the surrounding hills of the

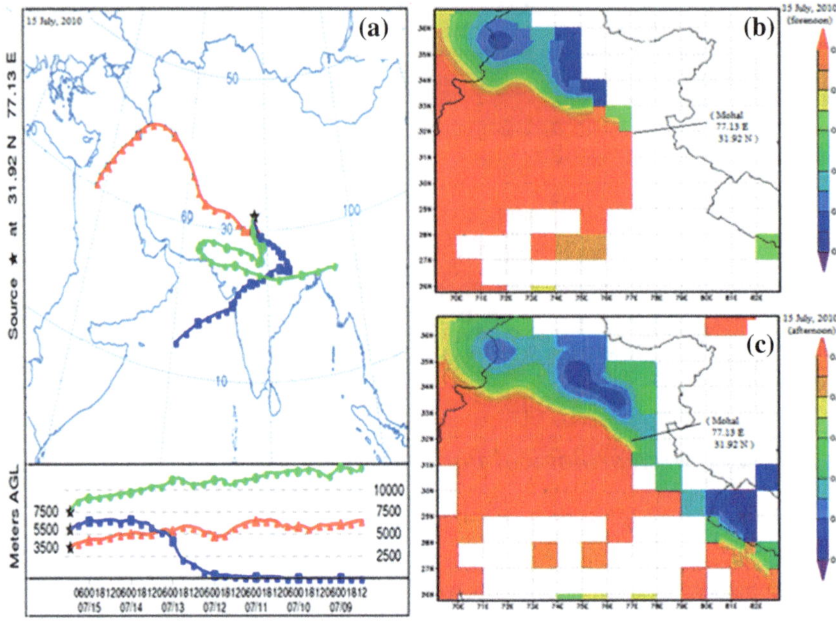

Fig. 4 Transported aerosols on July 15, 2010: **a** representing the back trajectory, **b** showing forenoon (Terra), and **c** afternoon (Aqua) AOD from MODIS satellite data

present experimental site located between 3,000 and 5,000 m (Sharma et al. 2009). The trajectories drawn at 5,500 and 7,500 m are influenced by the activities generating pollutants which are transported from the external sources outside the present study region—the Kullu valley.

The external sources of pollutants are sometimes from the other countries. The trajectories drawn at lower heights contribute to coarse size particles whereas at higher altitudes these contribute to the finer particles. A web based application known as GIOVANNI developed by Gooddard Earth Sciences Data and Information Science Centre (GESDISC) is used to establish relationship of our ground based measurements with the satellite based AOD at 550 nm. MODIS shows AOD for forenoon by using satellite Terra from the data product file MODO8_D3.051, while afternoon AOD was obtained using satellite Aqua from the data product file MYDO8_ D3.051. The products were MODIS Terra and Aqua Daily level-3 data (atmosphere daily global 1 × 1 degree) which were used in plotting the observation points over the land (website: http://www.disc.sci.gsfc.nasa.gov/Giovanni).

Based on the top four highest AOD values pinpointed during a recent past year 2010, it is observed that all these values ranged from 0.40 (June 19) to 0.50 (July 15) (Fig. 4b, c). When the back trajectory analysis in relation to MODIS data—Terra and Aqua for AOD is observed, it is found that the aerosol source regions during these 4 days were from the desert regions. The dust event of March 19,

2010, when AOD measured 0.44 through MWR, shows that all the three air masses at different altitudes were found to be a little bit influenced from the different directions.

However, when they reach covering half of the distance being covered during 7 days, they almost move together over the same places of the desert regions of the Middle East Countries and/or the Sahara Desert or over the places having more values of AOD ranging from 0.26 to 0.66 obtained through Terra and Aqua MODIS satellite data respectively. The other pollution episode, observed on April 6, 2010, shows 0.43 AOD from MWR while the regions through which the air masses move showed AOD in a range from 0.24 to 0.64 from MODIS—Terra and Aqua. In other words, it is made clear that these pollution episodes have been an outcome of external sources through air masses from the polluted regions. The event of pollution episode on June 19, 2010 also explains that the air mass at different three altitudes move from the three different directions. But the interesting part of these air masses before reaching to the present study region is that all these air masses move from the Sahara Desert. It is evident from these air masses that these passed through the highly polluted regions of Iran, Afghanistan and Pakistan.

3.5 AOD, Temperature Rise and Clear-Sky Shortwave Aerosol Radiative Forcing

During the observation period (2006–2010), the monthly 24-h average clear-sky shortwave aerosol radiative forcings (ARF) estimated at top of the atmosphere (TOA), surface (SFC), and atmosphere (ATM) were $+0.17 \pm 3.3$ W m^{-2}, -19.2 ± 2.1 W m^{-2} and $+19.4 \pm 3.7$ W m^{-2}, respectively. The clear-sky shortwave ARF varied from -6.4 to 6.1 W m^{-2}, -15 to -25.4 W m^{-2}, and $+15.1$ to $+29.0$ W m^{-2} at the TOA, SFC and ATM, respectively. The magnitude of ARF at the TOA can either be positive or negative which strongly depends on SSA (George 2001). The magnitude of TOA forcing is negative in pre-monsoon (-2.5 ± 2.0 W m^{-2}) and monsoon (-1.9 ± 1.3 W m^{-2}), whereas it is positive in post-monsoon ($+0.5 \pm 1.6$ W m^{-2}) and winter ($+4.0 \pm 1.7$ W m^{-2}). The seasonal maximum ARF at the surface occurred during pre-monsoon (-21.5 ± 2.1 W m^{-2}), while the minimum occurred during monsoon (-18.0 ± 1.3 W m^{-2}). During pre-monsoon, monsoon, post-monsoon and winter, the mean values of atmospheric forcing are estimated to be 19.0 ± 1.0 W m^{-2}, 16.0 ± 0.8 W m^{-2}, 19.9 ± 1.4 W m^{-2} and 23.0 ± 2.0 W m^{-2}, respectively. The highest value of atmospheric forcing in winter attributed to the absorptive properties of the aerosols. The aerosol forcing at the surface during pre-monsoon was about 1.2 times stronger than in winter. The considerable reduction in surface reaching solar radiation during pre-monsoon has increased the atmospheric forcing, which translates into a clear-sky atmospheric heating rate of 0.48 K day^{-1} in the 300 hPa thick aerosol layer. This strong response to aerosol forcing is attributed to enhanced loading of desert dust aerosols, which are mostly transported from the western desert regions. The atmospheric heating rate during pre-monsoon, monsoon, post-monsoon, and winter varied between 0.43–0.53, 0.42–0.49, 0.52–0.62, and 0.52–0.81 K day^{-1}, respectively. For comparison, Ganguly and Jayaraman (2006), Babu et al. (2007), Sreekanth et al. (2007), and Pathak et al. (2010) have reported an average range of heating rate for different seasons over Dibrugarh, Ahmedabad, Visakhapatnam, and Trivandrum to be 0.35–1.0, 0.60–1.13, 0.09–1.23, and 0.62–1.51 K day^{-1} respectively.

In the present study, the minimum value of AOD at 500 nm (AOD = 0.10) is obtained on 8 March 2007 and is referred as a low aerosol loading day. During pre-monsoon, the average forcing at the surface and in the atmosphere is found to change by −13.0 and +9.8 W m^{-2}, respectively when compared with the estimated value of a low aerosol loading day. This implies that during pre-monsoon, the reduction in surface reaching solar radiation increased by 152 % when is compared with the low aerosol loading day. At the same time, the atmospheric heating rate noted to be 1.8 times higher than that of low aerosol loading day. During post-monsoon, the average forcing at the surface and in the atmosphere is found to change by −10.9 and +10.7 W m^{-2}, respectively when compared with the value of a low aerosol loading day. It means that during post-monsoon, the reduction in surface reaching solar radiation increased by 128 % when it is compared with low aerosol loading day and the atmospheric heating rate touched 2.2 times higher when is compared with low aerosol loading day.

Largest annual-mean surface aerosol forcing is estimated to occur in 2006 and 2010 (Table 1). In contrast, the negative TOA aerosol forcing is estimated for year 2006 and 2010. The atmospheric forcing increases if aerosol forcing at the TOA approaches towards positive, while there is a large negative forcing at the surface level. This cause-effect relationship is due to absorption by absorbing aerosols (Pathak et al. 2010). It means that during 2007–2009, Mohal has been influenced due to a large concentration of absorbing aerosols as compared to their concentrations in 2006 and 2010.

During 2006–2010, the maximum and minimum values of AOD at 500 nm are obtained as 0.50 and 0.10 on 4 June 2009 and 8 March 2007, respectively. About 57 % decrease in the Ångström exponent and 500 % increase in the turbidity coefficient is observed on 4 June 2009 as compared to 8 March 2007, suggesting a large loading of coarse-mode size aerosols in the atmosphere. The possible

Table 1 Annual-mean clear sky aerosol radiative forcing: at the Earth's surface, TOA and in the atmosphere over Mohal

	Aerosol radiative forcing (W m^{-2})			
Year	TOA	Surface	Atmosphere	Heating rate (K day^{-1})
2006	−0.9	−19.5	18.5	0.52
2007	0.6	−18.5	19.1	0.54
2008	1.2	−18.9	20.1	0.56
2009	0.5	−19.1	19.6	0.55
2010	−0.6	−20.2	19.6	0.55

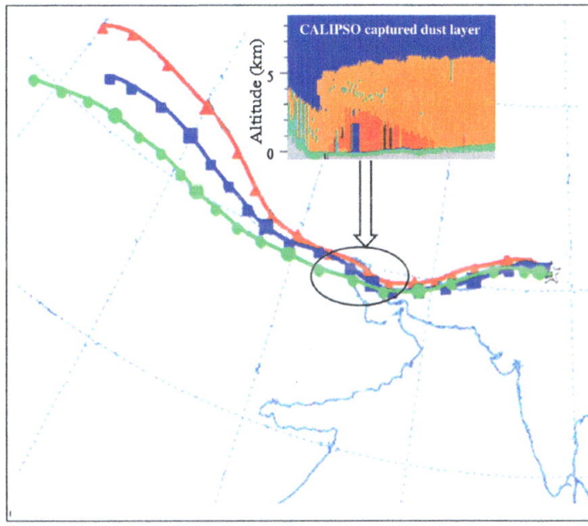

Fig. 5 Back trajectories ending at 06 UTC over Mohal drawn on highest AOD day at 500 nm, i.e., 4 June 2009, and CALIPSO derived vertical feature mask image (version: 3.01) showing large aerosol loading around 28.5°N, 50.0°E on 2 June 2009. The star denotes the location of Mohal

sources are investigated using the collective approach of CALIPSO and HYSPLIT model. The CALIPSO is a part of the A-Train constellation of satellites and has been making global measurement of aerosols and clouds since 13 June 2006 (Winker et al. 2009). The depolarization technique of the Cloud-Aerosol Lidar with Orthogonal Polarization (CALIOP) on board of CALIPSO satellite measurements is used to distinguish different variety of aerosols. After plotting back trajectories at different altitudinal heights, it was concluded that these air parcels originated in the Sahara Desert (Fig. 5). It was on 2 June 2009 when these air parcels passed through 28.5°N latitude, and 50.0°E longitude. At the same time, the CALIPSO pass was available which captured vertically an extended aerosol layer with a thickness of about 5 km (Fig. 5).

Moreover, the extinction-derived higher depolarization ratio (~0.4) profile derived from CALIOP suggested larger concentrations of dust particles. From the combined approach, it is known that the air-masses capture the dust while passing from around 28.5°N and 50.0°E and has increased the possibility to accumulate coarse-mode size particles over Mohal region. Figure 6 shows the influence of AOD on aerosol forcing estimated on 8 March 2007 and 4 June 2009. It can be seen from Fig. 6 that the enhanced loading of aerosols has a large impact with considerable reduction in surface reaching solar radiation. On 4 June 2009, the surface aerosol forcing is found to change by -14.7 W m^{-2} when is compared with 8 March 2007. Due to this large reduction in surface reaching solar radiation, the atmospheric heating touched 2.0 times higher on 4 June 2009 as compared to the value of 8 March 2007.

The notable reduction in surface reaching solar radiation estimated during a period of half decade (2006–2010) showed increasing atmospheric forcing, which translates into clear-sky atmospheric heating rate of 0.54 K day^{-1} in a 300 hPa thick aerosol layer with a month-to-month variations from 0.42 to 0.81 K day^{-1}. Such reduction in surface reaching solar radiation and increase in heating rate

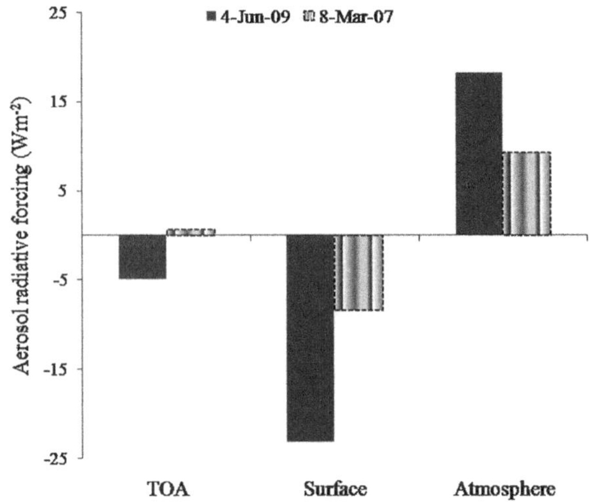

Fig. 6 Aerosol radiative forcing estimates on 8 March 2007 (a low aerosol loading day) and 4 June 2009 (a high aerosol loading day)

is responsible to cause changes in atmospheric thermal structure, synoptic and regional circulation systems, suppression of rainfall and less efficient removal of pollutants (Ramanathan and Feng 2009). The present study therefore reports the atmospheric forcing and associated heating rate, which directly supports the hypothesis of "Elevated Heat Pump", suggesting that absorbing aerosols over the northwestern Indian Himalaya may cause enhanced heating in the middle and upper troposphere, and may also lead to an early strengthening of the Indian monsoon rainfall (Lau et al. 2006; Lau and Kim 2006). This study suggests that while assessing the aerosol deposition (Ming et al. 2008; Qian et al. 2011) and aerosol-induced warming effects (Lau et al. 2010; Qian et al. 2011) on the glacier melting in the Himalayas, the seasonal variations in aerosol radiative forcing and solar heating rate over Mohal will be helpful in future for the radiative and climate impact assessments.

4 Conclusions

Aerosols have direct impact not only on the atmospheric visibility, air quality, formation of dew, mist and fog, etc. but also on the human health, plant life and more importantly on the climate. The study of aerosols in the Himalayan ecosystem is important because it will have long-term effects on temperature rise, shifting of vegetation and crops, glacier melting and human health in a wider altitudinal range from low to high. AOD values at shorter wavelengths are higher than at larger wavelengths indicating high anthropogenic pressure in the present study region which is increasing continuously. Afternoon AOD values are higher than forenoon AOD values being favoured by the strong convective activity during afternoon in presence of adequate solar flux. The monthly mean low Ångström exponent values noticed during pre-monsoon season indicate dominance of coarse size particles, while higher Ångström exponent values during post-monsoon season show dominance of fine size particles in our atmosphere. The seasonal aerosol forcing at the surface is estimated maximum during pre-monsoon, while the minimum occurred during monsoon. The surface aerosol forcing during pre-monsoon is found to be about 1.2 times stronger than winter. This study shows that during pre-monsoon, the reduction in the solar radiation reaching the surface increased by 152 % when is compared with the estimated value of a low aerosol loading day. Largest annual-mean surface aerosol forcing is estimated during 2010, while the minimum occurred during 2007. The present study estimates reduction in solar radiation arriving at the surface together with the atmospheric warming from 0.42 to 0.81 K day^{-1} with a mean value 0.54 K day^{-1} at 300 hPa thick aerosol layer.

Acknowledgments The authors are thankful to the Director, G. B. Pant Institute of Himalayan Environment and Development, Kosi-Katarmal, Almora, Uttarakhand for providing facilities in Himachal Unit of the Institute which could make this study possible. The authors also acknowledge with thanks to ISRO, Bangalore for providing financial assistance to the present project on 'aerosols climatology' under Aerosols Radiative Forcing over India (ARFI) programme of ISRO-GBP through Space Physics Laboratory, VSSC, Thiruvanthapuram, Kerala. Thanks are also due to NASA for providing MODIS and CALIPSO aerosol data and the NOAA Air Resources Laboratory for a provision of the HYsplit transport and dispersion model which are used in the present publication.

References

Ångström A (1961) Techniques of determining the turbidity of the atmosphere. Tellus 13:214–223

Babu SS, Moorthy KK, Satheesh SK (2007) Temporal heterogeneity in aerosol characteristics and the resulting radiative impacts at a tropical coastal station. Part 2: Direct short wave radiative forcing. Ann Geophys 25:2309–2320

Badrinath KVS, Kharol SK (2008) Studies on aerosols properties during ICARB-2006 campaign period at Hyderabad, India using ground-based measurements and satellite data. J Earth Syst Sci 117(S1):413–420

Beegum SN, Moorthy KK, Nair VS, Babu SS, Satheesh SK (2008) Characteristics of spectral aerosol optical depths over India during ICARB. J Earth Syst Sci 117:303–313

Bhuyan PK, Gogoi MM, Moorthy KK (2005) Spectral and temporal characteristics of aerosol optical depth over a wet tropical location in North East India. Adv Space Res 35:1423–1429

Chandrasekhar S (1950) Radiative Transfer. New York: Dover. ISBN 0-486-60590-6

Charlson RJ, Schwartz SE, Hales JM, Cess RD, Coakley JA, Hansen JE, Hofmann DJ (1992) Climate forcing by anthropogenic aerosols. Science 255:423–430

Draxler RR, Rolph GD (2010) Hybrid single-particle Lagrangian integrated trajectory model. Access via NOAA ARL READY website NOAA Air Resources Laboratory, Silver Spring, MD. http://ready.arl.noaa.gov/HYSPLIT.php

Dumka UC, Moorthy KK, Pant P, Hedge P, Sagar R, Pandey K (2008) Physical and optical characteristics of atmospheric aerosols during ICARB at Manora Peak, Nainital. a sparsely inhabited, high-altitude location in the Himalayas. J Earth Syst Sci 117(S1):399–405

Fochesatto GJ, Drobinski P, Flamant C, Guedalia D, Sarrat C, Flamant PH, Pelon J (2001) Evidence of dynamical coupling between the residual layer and the developing convective boundary layer. Boundary-Layer Meteorol 99:451–464

Fu Q, Liou KN (1992) On the correlated *k*-distribution method for radiative transfer in non-homogenous atmospheres. J Atmos Sci 49:2139–2156

Fu Q, Liou KN (1993) Parameterization of the radiative properties of cirrus clouds. J Atmos Sci 50:2008–2025

Ganguly D, Jayaraman A (2006) Physical and optical properties of aerosols over an urban location in western India: implications for shortwave radiative forcing. J Geophys Res, 111, D24–D207, doi:10.1029/ 2006JD007393

Ganguly D, Jayaraman A, Gadhavi H (2006) Physical and optical properties of aerosols over an urban location in western India: seasonal variabilities. J Geophys Res 111:D24206. doi:10.1029/2006JD007392

George JP (2001) Shortwave radiative forcing by mineral dust aerosols over Arabian Sea: a model study. Curr Sci 80:97–99

Gogoi MM, Bhuyan PK, Moorthy KK (2008) Estimation of the effect of long-range transport on seasonal variation of aerosols over northeastern India. Ann Geophys 26:1365–1377

Gogoi MM, Bhuyan PK, Moorthy KK (2009) An investigation of aerosol size distribution properties at Dibrugarh: North-Eastern India. Terrestrial Atmos Oceanic Sci 20:521–533

Guleria RP, Kuniyal JC, Rawat PS, Thakur HK, Sharma M, Sharma NL, Singh M, Chand K, Sharma P, Thakur AK, Dhyani PP, Bhuyan PK (2011a) Aerosols optical properties in dynamic atmosphere in the north western part of the Indian Himalaya: a comparative study from ground and satellite based observations. Atmos Res 101:726–738

Guleria RP, Kuniyal JC, Rawat PS, Sharma NL, Thakur HK, Dhyani PP, Singh M (2011b) The assessment of aerosol optical properties over Mohal in the northwestern Indian Himalaya using satellite and ground based measurements and an influence of aerosol transport on aerosol radiative forcing. Meteorol Atmos Phys 113 doi:10.1007/s00703-011-0149-5

Hansen ADA (2005) The Aethalometer Magee Scientific Company Berkeley, California, pp 1–207

Hansen ADA, Rosen H, Novakov T (1984) The aethalometer: an instrument for the real-time measurements of optical absorption by aerosol particles. Sci Total Environ 36:191–196

Hess M, Koepke P, Schult I (1998) Optical properties of aerosols and clouds: the software package OPAC. Bull Am Meteorol Soc 79:831–844

Holben BN, Tanre D, Smirnov A, Eck TF (2001) An emerging ground-based aerosol climatology: aerosol optical depth from AERONET. J Geophys Res 106:12067–12097

IPCC (2007) Climate change: the physical science basis. Cambridge University Press, Cambridge

Iqbal M (1983) An introduction to solar radiation. Academic Press, New York

Jin Z, Charlock TP, Rutledge K, Stamnes K, Wang Y (2006) Analytical solution of radiative transfer in the coupled atmosphere-ocean system with a rough surface. Appl Opt 45:7443–7455

Kuniyal JC, Rao PSP, Momin GA, Safai PD, Tiwari S, Ali K (2007) Trace gases behaviour in sensitive areas of the northwestern Himalaya-A case study of Kullu-Manali tourist complex, India. Indian J Radio Space Phys 36:197–203

Kuniyal JC, Thakur A, Thakur HK, Sharma S, Pant P, Rawat PS, Moorthy KK (2009) Aerosols optical depths at Mohal-Kullu in the northwestern Indian Himalayan high altitude station during ICARB. J Earth Syst Sci 118(1):41–48

Lau KM, Kim KM (2006) Observational relationships between aerosol and Asian monsoon rainfall, and circulation. Geophys Res Lett 33:L21810. doi:10.1029/2006GL027546

Lau KM, Kim MK, Kim KM (2006) Asian summer monsoon anomalies induced by aerosol direct forcing: the role of the Tibetan Plateau. Clim Dyn 26:855–864. doi:10.1007/s00382-006-0114-z

Lau KM, Kim MK, Kim KM, Lee WS (2010) Enhanced surface warming and accelerated 5 snow melt in the Himalayas and Tibetan Plateau induced by absorbing aerosols. Environ Res Lett 5:025204. doi:10.1088/1748-9326/5/2/025204

Liou KN (1973) Numerical experiment on Chandrasekhar's discrete ordinate—method for radiative transfer: applications to cloudy and haze atmosphere. J Atmos Sci 30:1303–1326

Liu SC, McKeen SA, Madronich S (1991) Effect of anthropogenic aerosols on biologically active ultraviolet radiation. Geophys Res Lett 8:2265–2268

Liu Z, Liu D, Huang J, Vaughan M, Uno I, Sugimoto N, Kittaka C, Trepte C, Wang Z, Hostetler C, Winker D (2008) Airborne dust distributions over the Tibetan Plateau and surrounding areas derived from the first year of CALIPSO lidar observations. Atmos Chem Phys 8:5045–5060

Ming J, Cachier H, Xiao C, Qin D, Kang S, Hou S, Xu J (2008) Black carbon record based on a shallow Himalayan ice core and its climatic implications. Atmos Chem Phys 8:1343–1352. doi:10.5194/acp-8-1343-2008

Moorthy KK, Nair PR, Murthy BVK (1991) Size distribution of coastal aerosols: effects of local sources and sinks. J Appl Meteorol 30:844–852

Moorthy KK, Nair PR, Murthy BVK, Satheesh SK (1996) Time evolution of the optical effects and aerosol characteristics of Mt. Pinatubo origin from ground-based observations. J Atmos Terr Phys 58:1101–1116

Moorthy KK, Satheesh SK, Murthy BVK (1998) Characteristics of spectral optical depths and size distributions of aerosols over tropical oceanic regions. J Atmos Terr Phys 60:981–992

Moorthy KK, Niranjan K, Narasimhamurthy B, Agashe VV, Murthy BVK (1999) Aerosol climatology over India, ISRO-GBP MWR network and data base. Scientific report, SR-03-99. Indian Space Research Organisation, Bangalore

Moorthy KK, Babu SS, Satheesh SK (2005) Aerosol characteristics and radiative impacts over the Arabian Sea during intermonsoon season: results from ARMEX field campaign. J Atmos Sci 62:192–206

Nair VS, Moorthy KK, Alappattu DP, Kunhikrishnan PK, George S, Nair PR, Babu SS (2007) Wintertime aerosol characteristics over the Indo-Gangetic Plain (IGP): impacts of local boundary layer processes and long-range transport. J Geophys Res 112:D13205. doi:10.1029/2006JD008099

Pant P, Hegde P, Dumka UC, Sagar R, Satheesh SK, Moorthy KK, Saha AM, Srivastava MK (2006) Aerosol characteristics at a high altitude location in Central Himalayas: optical properties and radiative forcing. J Geophys Res 111:D17206. doi:10.1029/2005JD006768

Pathak B, Kalita G, Bhuyan K, Bhuyan PK, Moorthy KK (2010) Aerosol temporal characteristics and its impact on shortwave radiative forcing at a location in the northeast of India. J Geophys Res 115:D19204. doi:10.1029/2009JD013462

Pinker RT, Pandithurai G, Holben BN, Dubovik O, Aro TO (2001) A dust out break episode in sub-sahel West Africa. J Geophys Res 106(22):923–930

Qian Y, Flanner MG, Leung LR, Wang W (2011) Sensitivity studies on the impacts of Tibetan Plateau snowpack pollution on the Asian hydrological cycle and monsoon climate. Atmos Chem Phys 11:1929–1948. doi:10.5194/acp-11-1929-2011

Raju MP, Safai PD, Rao PSP, Devara PCS, Budhavant KB (2011) Seasonal characteristics of black carbon aerosols over a high altitude station in southwest India. Atmos Res 100:103–110

Ramanathan V, Feng Y (2009) Air pollution, greenhouse gases and climate change: global and regional perspectives. Atmos Environ 43:37–50

Ricchiazzi P, Yang SR, Gautier C, Sowle D (1998) SBDART: a research and teaching software tool for plane-parallel radiative transfer in the Earth's atmosphere. Bull Am Meteorol Soc 79(10):2101–2114

Satheesh SK, Moorthy KK, Das I (2001) Aerosol spectral optical depths over the Bay of Bengal, Arabian Sea and Indian Ocean. Curr Sci 81(12):1617–1625

Sharma NL, Kuniyal JC, Singh M, Negi AK, Singh K, Sharma P (2009) Number concentration characteristics of ultrafine aerosols (atmospheric nanoparticles/aitken nuclei) during 2008 over western Himalayan region, Kullu-Manali, India. Indian J Radio Space Phys 38:326–337

Shaw GE, Reagan JA, Herman BM (1973) Investigations of atmospheric extinction using direct solar radiation measurement made with a Multiple Wavelength Radiometer. J Appl Meteorol 12:374–380

Singh M, Singh D, Pant P (2008) Aerosols characteristics at Patiala during ICARB-2006. J Earth Syst Sci 117(S1):407–411

Singh S, Soni K, Bano T, Tanwar RS, Nath S, Arya BC (2010) Clear-sky direct aerosol radiative forcing variations over mega-city Delhi. Ann Geophys 28:1157–1166

Sreekanth V, Niranjan K, Madhavan BL (2007) Radiative forcing of black carbon over eastern India. Geophys Res Lett 34:L17818. doi:10.1029/2007GL030377

Stamnes K, Tay SC, Wiscombe WJ, Jayweera K (1988) Numerically stable algorithm for discrete-ordinate-method radiative transfer in multiple scattering and emitted media. Appl Opt 27:2502–2509

Stull RB (1988) An introduction to boundary layer meteorology. Kluwer Academic Publishers, Dordrecht 670

Winker DM, Vaughan MA, Omar A, Hu Y, Powell KA, Liu Z, Hunt WH, Young SA (2009) Overview of the CALIPSO mission and CALIOP data processing algorithms. J Atmos Oceanic Technol 26:2310–2323

Measurement of Atmospheric Carbon Dioxide Levels at Dokriani Bamak, Garhwal Himalaya, India

Ashish Anthwal, Varun Joshi, Suresh C. Joshi and Kireet Kumar

Abstract Increase in surface temperature at global scale has already affected a diverse set of physical and biological systems in many parts of the world and if it increases at this rapid rate then the condition would be worst one could have ever thought off. Garhwal Himalaya, major part of the great Himalayan mountainous system is also much sensitive and vulnerable to the local, regional and global changing climate. Due to large altitudinal gradient, varied climatic conditions and diverse set of floral and faunal composition, the impact of climate change seems to be much perceptible in coming future. Natural ecosystems at high elevations are much more sensitive to the climatic variations or global warming then the managed systems. This paper highlights measurement of atmospheric Carbon dioxide at Dokriani Bamak, Uttarkashi District, Uttarakhand. Concentration of CO_2 averaged 383.5 ± 2.12 ppm in 2005. Daily variations of CO_2 values showed minimum during the daytime (376.5 ppm) and peaked in the evening (393.8 ppm). At monthly intervals, the CO_2 values varied from 381.9 ± 3.70 (May) to 385.52 ± 7.05 ppm (August). Average temperature recorded during the year was 4.7 °C and during the growing season (May–October 2005) was 6.8 °C. Although phenology is significant in controlling CO_2 levels, short-term changes

A. Anthwal
Public Works Department, Dehradun, Uttarakhand, India

V. Joshi (✉)
University School of Environment Management,
GGS Indraprastha University, New Delhi, India
e-mail: varunj63@gmail.com

S.C. Joshi
G. B. Pant Institute of Himalayan Environment and Development,
Sikkim Unit, Pangthang, Sikkim, India

K. Kumar
G. B. Pant Institute of Himalayan Environment and Development,
Kosi-Katarmal, Almora, Uttarakhand, India

R. Joshi et al. (eds.), *Dynamics of Climate Change and Water Resources of Northwestern Himalaya*, Society of Earth Scientists Series,
DOI 10.1007/978-3-319-13743-8_10

cannot be explained without the anthropogenic perturbations. The CO_2 concentration in Dokriani Bamak (383.5 ppm) was higher and comparable with those of other major monitoring locations around the world.

Keywords Atmospheric carbon dioxide · Diurnal · Phenology · Climate change · Dokriani Bamak · Garhwal Himalaya

1 Introduction

Carbon dioxide is a trace gas in the earth's atmosphere of which exchange occurs between the major environmental reservoirs such as the oceans and the biosphere. Being the most abundant greenhouse gas in the atmosphere (besides water vapour) atmospheric CO_2 contributes most significantly to global climate change (Bolin et al. 1986; Houghton et al. 1996). The great industrial demands for energy promoted to release vast quantities of carbon dioxide into the earth's atmosphere since the beginning of the industrial revolution. Monitoring of atmospheric CO_2, as conducted constantly since the late 19th century, showed a steady and continuous increase in its concentrations. Because of the increasing human activities, its atmospheric concentration increased from 280 ppm in pre-industrial revolution to current level of 380 ppm (IPCC 2007). According to IPCC (2001), atmospheric CO_2 concentrations increased by 31 % over the last 250 years. The average increase rate of CO_2 was maintained at 1.4 ppm year^{-1} for the period 1960–2005 (IPCC 2007). The rate of its increase in the last 10 years (1995–2005) is estimated to be 1.9 ppm year^{-1} to show the highest growth rate since its direct measurements from 1950s (IPCC 2007).

Carbon dioxide levels recorded at various locations around the world consistently show a direct link with fossil fuel combustion (Denning et al. 1995; Colombo et al. 2000). Other sources of atmospheric CO_2 include plants, animals, microbial respiration, ocean emission, and land use change. As such, fossil fuel combustion and cement production have increased carbon dioxide emissions by 70 % in the last 30 years (Prentice et al. 2001; Marland et al. 2006). As the increase in atmospheric greenhouse gas concentrations is the main cause of global warming, it is predicted to affect trend of climate change in both regional and global scales. Detailed information concerning the source/sink of greenhouse gases and their emission strengths has been one of the major goals of global climate study. Because the distribution of CO_2 is subject to geographical and temporal variations (Keeling 1961; Pales and Keeling 1965; Inoue and Matseuda 1996), model predictions with a relatively wide geographical coverage were not necessarily useful to accurately quantify CO_2 exchange on a global scale (Massarie and Tans 1995; Keeling et al. 1995). By considering such limitations, numerous attempts have been made to build a database of CO_2 to cover diverse environmental conditions (Levin 1987; Levin et al. 1995; Schmidt et al. 1996). For instance, a continuous measurement of CO_2 (and the related isotopic carbon ratios) has been reported in the Krakow region, Poland (Kuc 1991) or in the K-puszta region of Hungary (Haszpra 1995). However, despite the importance of CO_2 data acquisition, relatively little is known about its distribution on hilly areas of the world. This paper reports the results of CO_2 measurements made in the atmosphere of Dokriani Bamak lying in the high altitude region of the Northwest Himalaya, India. The purpose of this study is to primarily evaluate the carbon dioxide concentration levels and its temporal variabilities in the hilly region of Garhwal Himalaya at Dokriani Bamak, India. The results of this study will provide some insights into the environmental behaviour of CO_2 in mountainous environs.

2 Materials and Methods

The concentration data of CO_2 were collected using infrared CO_2 gas analyzer (LI- 820, LI-COR, USA). Air was drawn at a flow rate of 1 L min^{-1} through air filter (Balston 25 μm) attached to non-CO_2 absorbing Teflon tubing into the gas analyzer. Sampling interval was set to 30 s, and the data were recorded every 2 h by a data logger (LI-1400, LI-COR, USA). The incoming air was passed through a column of magnesium perchlorate to eliminate the possible interference due to water vapour. The CO_2 gas analyzer was calibrated prior to measurement by checking span and zero values with CO_2 calibrant gas (505 ppm). The measurement range of CO_2 by the NDIR analyzer was 0–1,000 ppm with an accuracy of <2.5 % and a total drift of <0.4 ppm/°C at 370 ppm. The CO_2 values recorded at two hourly intervals were converted into daily or monthly for the analysis of its temporal variabilities at different intervals.

3 Results and Discussion

3.1 General Pattern of Temperature and CO_2 Distribution at Study Site

The present study was conducted at Dokriani Bamak, a high altitude area situated in Uttarkashi district of Uttarakhand Himalayan region in India (Fig. 1).

Carbon dioxide concentrations were recorded at the Base camp of Dokriani Bamak (altitude of 3,600 m above mean sea level, Latitude 30° 50′–30° 52′N, Longitude 78° 47′–78° 50′E). The study area is accessible only during the summers i.e. May to November. During rest of the months the area is inaccessible due to heavy snow. So the temperature data were only measured during the growing season. Average temperature recorded during the year was 4.7 °C and during the growing season (May–October 2005) was 6.8 °C. Concentrations of CO_2 in air were monitored at 1.5 m above the ground at two hourly intervals (for up to 9 h: 0600 to 2200 h (local time) throughout the study period (May to

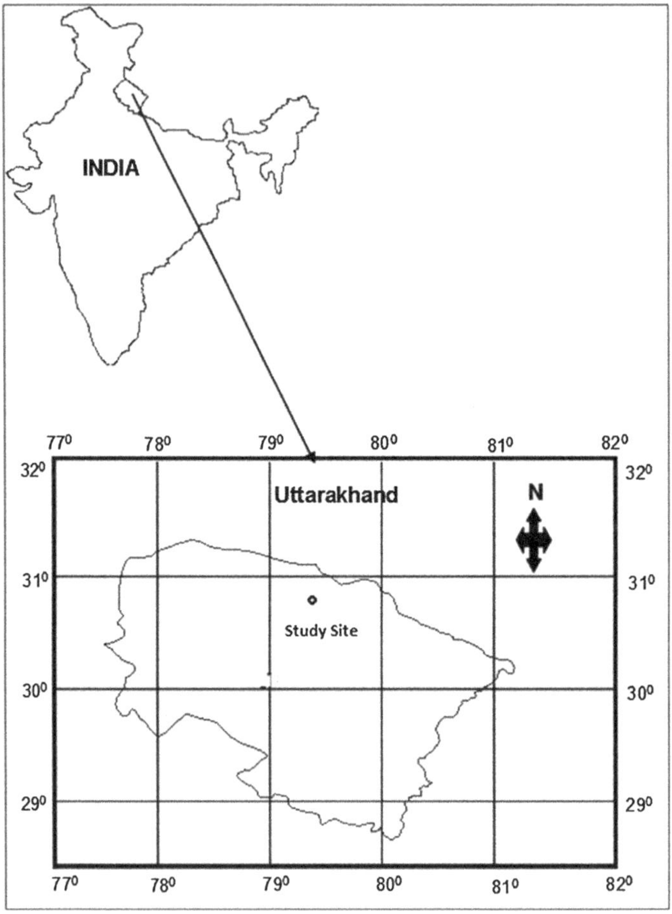

Fig. 1 Location map of the study site

November 2005 except September 2005). The concentration data of atmospheric CO_2 have been collected continuously from various locations in the world since its measurements at Antarctica and Mauna Loa observatory (Hawaii) in 1958. In India, CO_2 was monitored for the first time in the air and soil layers near the ground in the year 1941 (Mishra 1950): The study was conducted to measure CO_2 in the open as well as crop fields. The longest monitoring of atmospheric CO_2 in India was first made at Cape Rama, a maritime site located in the west coast of India for a 10 year period (1993 and 2002: Bhattacharya et al. 1997). The results of this study are the first attempt to continuously measure atmospheric carbon dioxide in the hilly region of Dokriani Bamak, Uttarkashi, Uttarakhand. In Table 1, the basic statistical parameters of CO_2 data monitored in this study are presented. The overall mean value of CO_2 measured in this study was 383.5 ± 2.8 ppm.

Table 1 Statistical summary of atmospheric CO_2 recorded at Dokriani Bamak, Uttarkashi

Months	Mean	Median	SD	Minimum	Maximum	N	SE	RA
May	381.85	381.00	3.7	378.41	388.13	9	1.23	9.72
Jun	385.95	381.95	8.6	377.64	400.61	9	2.86	22.97
Jul	384.44	385.09	7.5	373.85	394.18	9	2.49	20.33
Aug	385.51	385.49	7.0	373.69	395.49	9	2.35	21.8
Oct	380.61	379.85	4.2	375.01	387.32	9	1.41	12.31
Nov	382.82	381.01	5.1	376.96	390.48	9	1.69	13.52

3.2 *Diurnal Variation in Carbon Dioxide Levels*

To assess the diurnal variation of CO_2, the data obtained above the ground at every 2 h intervals are plotted and examined in many respects. In Fig. 2a, the CO_2 concentrations for each 2 h interval are compared diurnally using all data sets. The CO_2 levels at this site maintained a diurnal pattern that is consistent enough to show the highest values (393.8 ppm) in the night time (2200 h) and lowest values (376.5 ppm) in the afternoon (1200 h) with the relative amplitude of 4.5 %. Figure 2b depicts the diurnal pattern of CO_2 between different months of the year 2005. The differences in hourly CO_2 concentration levels varied significantly between minimum (373.7 ppm in August) and maximum values (400.6 ppm in June) across the months. If the strengths of diurnal variation were compared by

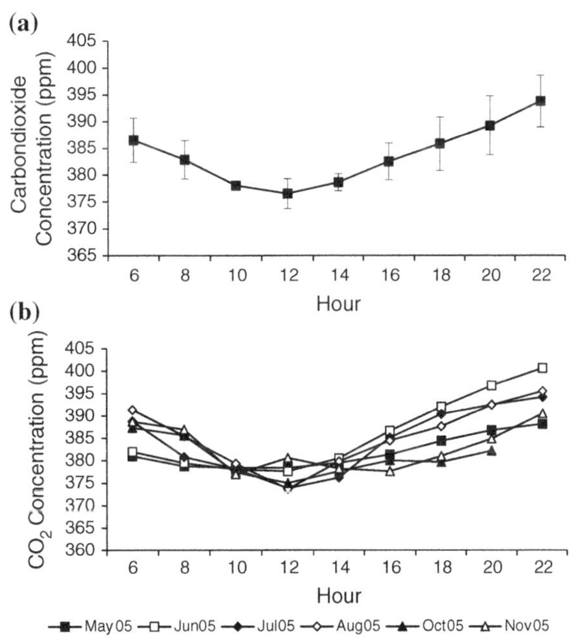

Fig. 2 Diurnal variation in the atmospheric CO_2 using all hourly measurements for **a** all data sets and **b** monthly data sets of CO_2

relative amplitude (RA) values between different months, the RA values ranged from 2.55 (May) to 5.95 % (June). The diurnal cycle of CO_2 is generally known to exhibit a maximum at night (or morning) and a minimum during the daytime (Schnell et al. 1981; Baez et al. 1988; Yi et al. 2001). A nighttime maximum of CO_2 in rural areas has been attributed to respiration by plants (and animals) and its emissions from soils. In contrast, daytime minimum is explained by photosynthesis (Spittlehouse and Ripley 1977; Baez et al. 1988; Nasrallah et al. 2003). Such phenomenon can also be explained partially by the changes in meteorological conditions, as the height of the mixing layer increases under stronger solar radiation (Aikawa et al. 1995).

Many previous studies based on long-term monitoring of CO_2 also indicated that CO_2 fluctuates both diurnally and seasonally (Woodwell 1978; Keeling et al. 1984; Fung et al. 1987). The average relative amplitude in this study was found as 6.95 % between the daytime drawdown and night time buildups of CO_2. This RA value is smaller than those measured in the Savannah regions (21.6 %) and tropical rain forests (25.4 %) (Schnell et al. 1981), urban area of Basel city (Switzerland) with 15.5 % (Vogt et al. 2006), and urban area of Chicago with RA values of 9.02 % (Grimmond et al. 2002). The RA value of the present study is however comparable to that measured from four different sites in Phoenix (2.85–7.86 %: Day et al. 2002). Differences in the magnitude of RA values may be ascribable to such factors as the strength of biospheric photosynthesis, respiration, mixing conditions, and emissions from anthropogenic sources (Pales and Keeling 1965; Inoue and Matsueda 1996). Considering the magnitude of diurnal fluctuations in the study area, such variability CO_2 may have significant implications on the vegetation of the region due to its impact on the plant photosynthesis (Veste and Herppich 1995).

4 Comparison with Previous Studies of CO_2 Concentration

In an attempt to understand the factors controlling the distribution of CO_2 under various environmental conditions, we examined our monitoring data obtained from the mountainous area of Garhwal Himalaya, India with those reported from other parts of the world. Table 2 summarizes the yearly carbon dioxide values, measurements conditions, detection method, and amplitude of CO_2 data for all comparable data sets. For this comparative analysis, all the reference data were basically taken from the data sets of year 2006 from the WMO global atmosphere watch, world data centre for greenhouse gases (WDCGG).

The annual mean concentrations of CO_2 for all the recording stations except Romania (368.3 ppm) were well above the global background concentrations of CO_2 (380 ppm). Figure 3 depicts the absolute concentrations and relative amplitude of CO_2 measured from all stations examined for comparative purposes. The mean CO_2 concentration for India, Himalaya during the study year was higher (383.5 ppm), when compared to the stationary stations in Australia (379.1 ppm),

Table 2 Comparison of the mean CO_2 concentration measured from all different stations around the world

City/station	Recording year	Method	Sampling type	No of sampling months	Average	SE	Amplitude	Relative amplitude
Australia (Cape Fergusson)	2006	GC (FID)	Flask	12	379.1	0.2	2.2	0.59
Austria (Sonnblick)	2006	NDIR	Continuous	9	381.71	1.8	13.0	3.41
Finland (Pallas-Sammaltunturi)	2006	NDIR	Continuous	12	384.38	1.8	18.0	4.68
Germany (Deuselbach)	2003	GC (FID)	Continuous	11	386.15	1.7	15.0	3.88
Israel (Sede Boker)	2006	NDIR	Flask	12	383.57	1.1	11.0	2.87
Kazakhstan (Sary Taukum)	2006	NDIR	Flask	12	385.35	1.8	16.0	4.15
Mongolia (Ulaan Uul)	2006	NDIR	Flask	12	383.66	1.6	17.0	4.43
Kenya (Mt. Kenya)	2006	NDIR	Flask	7	379.63	0.5	3.3	0.86
Norway (Zeppelinfjellet)	2006	NDIR	Continuous	12	383.07	1.6	15.0	3.92
Portugal (Terceira Island)	2007	NDIR	Flask	11	383.06	1.2	12.0	3.13
Russia (Teriberka)	2006	NDIR	Flask	11	384.63	1.6	17.6	4.58
Seychelles (Mahe Island)	2006	NDIR	Flask	12	380.42	0.3	3.0	0.79
USA (Southern Great Plains)	2006	NDIR	Flask	12	384.98	1.1	12.0	3.12
Romania (Fundata)	2005	NDIR	Continuous	12	368.32	4.3	54.1	14.69
India (Dokriani bamak)	2005	NDIR	Continuous	6	383.5	0.7	17.2	4.49

Source WMO global atmosphere watch, World data centre for greenhouse gases (WDCGG)

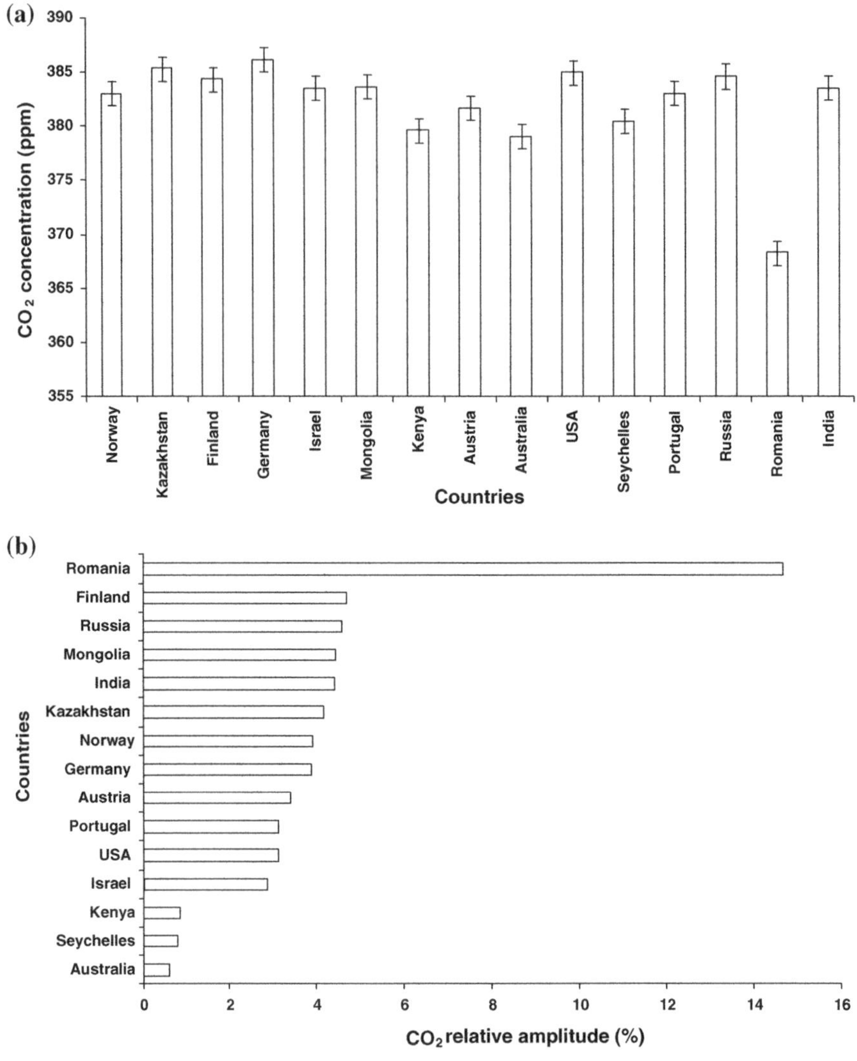

Fig. 3 A comparison of **a** absolute concentration (ppm) and **b** relative amplitude (%) of CO_2

Norway (383.07 ppm), Austria (381.7 ppm), Mt. Kenya (379.6 ppm) and Romania (368.3 ppm) stations. The relative amplitude of the CO_2 values for all the comparative data can be estimated as the difference between the maximum and minimum values (amplitude) over mean. The relative amplitude of our study site was found to be more or less similar (4.50 %) to the RA recorded at Finland (4.68 %), Mongolia (4.43 %), Russia (4.58 %) and Kazakhstan (4.15 %).

Figure 4 shows the comparison of the monthly mean values of CO_2 measured at different stations over the globe. Pallas-Sammaltunturi (Finland) and Mt. Kenya

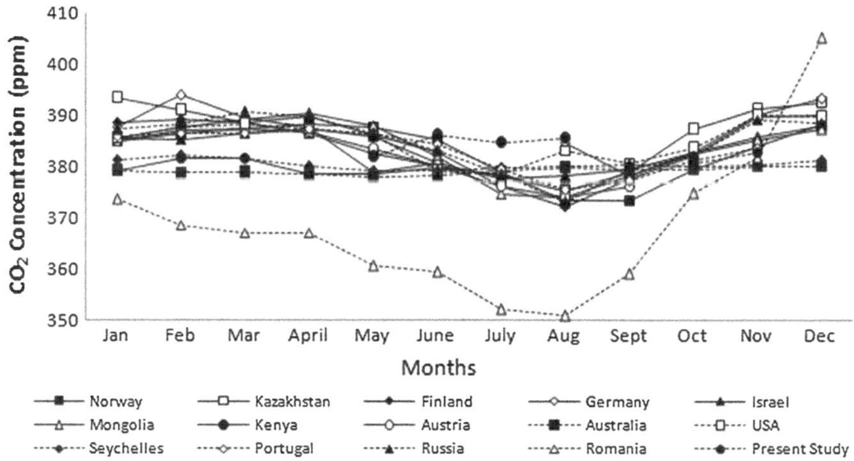

Fig. 4 Month- to- month variation of CO_2 in all stations selected for comparison

(Kenya) represent the global CO_2 concentration sites, whereas all the other stations for regional CO_2 concentration. Among all the sites shown in the Table 2, a number of stations including Sonnblick, Deuselbach, Fundata, Ulaan Uul, Sary taukum, etc. represent mountainous sites. All of these stations are stationary, while they are free from direct effect of any known anthropogenic sources. The annual mean CO_2 concentration for mountainous sites, if derived using all those data sets, was much lower or similar (384.2 ppm, range: 381.7–386.2 ppm) than that of our study (383.5 ppm). If the relative amplitude values are compared between all mountainous sites (3.41–4.43 %), their values are quite analogous to our results (4.50 %).

5 Conclusion

In the present study, the temporal variations in the atmospheric carbon dioxide in the mountainous area of Dokriani bamak were investigated using the data sets collected from May to Nov 2005 except September. The diurnal variation of CO_2 was characterized by relative enhancement in the night. As the green plants intensively absorb atmospheric CO_2 (through photosynthesis), the concentrations of CO_2 are maintained in the least level during the daytime. When the diurnal variations are assessed across different months, the patterns confirmed the combined effect of biogenic and meteorological factors. It should be stressed here that the mean carbon dioxide concentration during the growing season in Dokriani bamak were higher (383.5 ppm) than the global mean atmospheric CO_2 value of around 380 ppm. The present work is the first preliminary report covering continuous monitoring of CO_2 in the mountainous region of Garhwal Himalaya, India. According to our analysis, it may be important to explain the possible cause of the high CO_2 levels in this clean area. As the troposphere baseline data of CO_2 concentration were not measured over the Himalayan region previously, precise measurements of atmospheric CO_2 are needed for an extended period. Such efforts can offer more insights into the factors governing the CO_2 concentration under diverse environmental settings.

Acknowledgments Authors are thankful to the Director, G.B. Pant Institute of Himalayan Environment and Development for providing necessary facilities. Financial support provided by DST for purchasing the instrument is greatfully acknowledged.

References

Aikawa M, Yoshikawa K, Tomida M, Aotsuka F, Haraguchi H (1995) Continuous monitoring of the carbon dioxide concentration in the urban atmosphere of Nagoya, 1991–1993. Anal Sci 11:357–362

Baez A, Reyes M, Rosas I, Mosiño P (1988) CO_2 concentrations in the highly polluted atmosphere of Mexico City. Atmosfera 1:87–98

Bhattacharya SK, Jani RA, Borole DV, Francey RJ, Masarie KA (1997) Atmospheric carbon dioxide and other trace gases in a tropical Indian station. In: Gröning M, Gibert-Massault E (eds) First research coordination meeting, coordinated research programme on isotope-aided studies of atmospheric carbon dioxide and other greenhouse gases: report, Vienna, Austria. IAEA Isotope Hydrology Section, Vienna

Bolin B, Doos BR, Jager J, Warrick RA (1986) The greenhouse effect, climatic change, and ecosystems: SCOPE 29. Wiley, New York

Colombo T, Santaguida R, Capasso A, Calzolari F, Evangelisti F, Bonasoni P (2000) Biospheric influence on carbon dioxide measurements in Italy. Atmos Environ 34:4963–4969

Day TA, Gober P, Xiong FS, Wentz EA (2002) Temporal patterns in near-surface CO_2 concentrations over contrasting vegetation types in the Phoenix metropolitan area. Agric For Meteorol 110:229–245

Denning AS, Fung IY, Randall D (1995) Latitudinal gradient of atmospheric CO_2 due to a seasonal exchange with land biota. Nature 376:240–243

Fung IY, Tucker CJ, Prentis KC (1987) Application of advanced very high resolution radiometer vegetation index to study atmosphere-biosphere exchange of CO_2. J Geophys Res 92:2999–3015

Grimmond CSB, King TS, Cropley FD, Nowak DJ, Souch C (2002) Local-scale fluxes of carbon dioxide in urban environments: methodological challenges and results from Chicago. Environ Pollut 116:243–254

Haszpra L (1995) Carbon dioxide concentration measurements at a rural site in Hungary. Tellus 47B:17–22

Houghton JT, Meira Filho LG, Callander BA, Harris N, Kattenberg A, Maskell K (eds) (1996) Climate Change 1995. The science of climate change. Contribution of working group 1 to the second assessment report of the Intergovernmental panel on climate change. Cambridge University Press, Cambridge

Inoue HY, Matsueda H (1996) Variations in atmospheric CO_2 at the Meteorological Research Institute, Tsukuba, Japan. J Atmos Chem 23:137–161

Intergovernmental Panel on Climate Change (IPCC), Climate Change (2001) Radiative forcing of climate change: the scientific basis. Cambridge University Press, UK, p 892

Intergovernmental Panel on Climate Change (IPCC), Climate Change (2007) Synthesis report. In: Pachauri RK, Reisinger A (eds) Contribution of Working Groups I, II and III to the fourth assessment report of the Intergovernmental Panel on Climate Change Core Writing Team. IPCC, Geneva

Keeling CD (1961) The concentration and isotopic abundances of carbon dioxide in rural and marine air. Geochim Cosmochim Acta 24:277–298

Keeling CD, Carter AF, Mook WG (1984) Seasonal, latitudinal, and secular variations in the abundance and Isotopic ratios of atmospheric carbon dioxide: results from oceanographic cruises in the Tropical Pacific Ocean. J Geophys Res 89:4615–4628

Keeling CD, Whorf TP, Wahlen M, van der Plicht J (1995) Interannual extremes in the rate of rise of atmospheric carbon dioxide since 1980. Nature 375:666–670

Kuc T (1991) Concentration and carbon isotopic composition of atmospheric CO_2 in southern Poland. Tellus 43B:373–378

Levin I (1987) Atmospheric CO_2 in continental Europe—an alternative approach to clean air CO_2 data. Tellus 39B:21–28

Levin I, Graul R, Trivett NBA (1995) Long term observation of atmospheric CO_2 and carbon isotopes at continental sites in Germany. Tellus 47B:23–24

Marland G, Boden TA, Andres RJ (2006) Global, regional, and national CO_2 emissions. In: Trends: a compendium of data on global change. Carbon Dioxide Information Analysis Center, Oak Ridge National Laboratory, US Department of Energy, Oak Ridge, TN. Available online at http://cdiac.ornl.gov/trends/emis/em_cont.html

Masarie KA, Tans PP (1995) Extension and integration of atmospheric carbon dioxide data into a globally consistent measurement record. J Geophys Res 100:11593–11610

Misra RK (1950) Studies on the carbon dioxide factor in the air and soil layers near the ground. Indian J Meteorol Geophys 1(4):275–286

Nasrallah HA, Balling RC, Madi SM, Al-Ansari L (2003) Temporal variations in atmospheric CO_2 concentrations in Kuwait City, Kuwait with comparisons to Phoenix, Arizona, USA. Environ Pollut 121:301–305

Pales JC, Keeling CD (1965) The concentration of atmospheric carbon dioxide in Hawaii. J Geophys Res 24:6053–6076

Prentice IC, Farquhar GD, Fasham MJR (2001) The carbon cycle and atmospheric carbon dioxide. In: Houghton JT et al (eds) Climate change 2001: the scientific basis. Cambridge University Press, Cambridge, pp 183–237

Schmidt M, Graul R, Sartorius H, Levin I (1996) Carbon dioxide and methane in continental Europe: a climatology, and radon-based emission estimates. Tellus 48B:457–473

Schnell RC, Odh SA, Njau LN (1981) Carbon dioxide measurements in tropical east African biomes. J Geophys Res 86:5364–5372

Spittlehouse DL, Ripley EA (1977) Carbon dioxide concentration over a native grassland in Saskatchewan. Tellus 29:54–65

Veste M, Herppich WB (1995) Influence of diurnal and seasonal fluctuations in atmospheric CO_2 concentration on the net CO_2 exchange of poplar trees. Photosynthetica 31:371–378

Vogt R, Christen A, Rotach MW, Roth M, Satyanarayana ANV (2006) Temporal dynamics of CO_2 fluxes and profiles over a Central European city. Theoret Appl Climatol 84:117–126

WMO global atmosphere watch, World data centre for greenhouse gases (WDCGG) (2008) http://gaw.kishou.go.jp/cgi-bin/wdcgg/catalogue.cgi

Woodwell GM (1978) The carbon dioxide question. Sci Am 238(1):34–43

Yi C, Davis KJ, Berger BW (2001) Long-term observations of the dynamics of the continental planetary boundary layer. J Atmos Sci 58:1288–1299

Part III
Consequences of Changes and Flow Regime

Hydrological Management of Glacial and Non-glacial River Systems

Kireet Kumar, S. Joshi, V. Adhikari, H. Sharma and T. Pande

Abstract The present study investigates the specific features of the hydrologic behavior of glacial and non-glacial river systems. Glacial river systems are characterized as high energy landforms with less biotic activities, whereas the non-glacial river systems have gentle slopes and more intensive biotic activities. Glacial basins generate large flow variations on seasonal basis, but the mean daily discharge in different months is less variable. The role of frozen storage and difference in time of concentration is significant for maintaining the flow in compound glaciers. Non-glacial systems are more precipitation dependent and catchment characteristics play an important role in time of concentration. Headwaters of non glacial rivers have time lag of only few days and flow is not sustained after rains as reflected by the high discharge ratios. Flow duration curves with gentle slope of glacial basin indicate influence of glacier storage and release of water from different zones of the glacier which sustain the flow in the stream. The flow duration curves of non-glacial basin for different monsoon months describe the changing flow pattern. A large part of the year falls under dry weather and low flow conditions and the flow generated by the rain recedes very fast. The rate of increase and decrease of discharge with time (dry and wet season) depends upon type of source, storage characteristics of the aquifer and catchment area.

Keywords Hydrological management · Glacial river systems · Non-glacial river systems · Discharge · Glacier

K. Kumar (✉) · S. Joshi · V. Adhikari · H. Sharma · T. Pande
G.B. Pant Institute of Himalayan Environment and Development, Kosi-Katarmal, Almora, Uttarakhand, India
e-mail: kireetkumar@yahoo.com

R. Joshi et al. (eds.), *Dynamics of Climate Change and Water Resources of Northwestern Himalaya*, Society of Earth Scientists Series,
DOI 10.1007/978-3-319-13743-8_11

1 Introduction

Glaciers are important components of earth system; it controls the river hydrology of the mountainous area and polar region. The Himalaya constitutes one of the most important glacier systems of the world. In the Himalaya about 1,400 km^3 of snow and ice locked, and it spread over nearly 33,200 km^2 area in higher altitude above the 4,300–5,800 m, there are 9,000 glaciers in the Himalaya, this area is considered geodynamically active and it is prone to violent crustal movement causing seismicity (Valdiya 1998). The Ganga basin has 968 glaciers with a total glacier cover of 2,857 km^2 and total ice volume of 209.37 km^3 (Raina and Srivastava 2008). Himalayan glaciers are important because they are the major factors in controlling the climate in Himalayan regions and form major source of water in glacial rivers.

The Himalayan proglacial streams carry good portion of the total summer river flow, which is derived from snow and glacier ice melt (Bruijnzeel and Bremmer 1989). During the ablation season, melting of snow and ice at the glacier surface is the major source of glacier discharge, which in turn is related to the radiative energy input (Collins and Hasnain 1994). Consequently, the variations in air temperature and radiation influence ablation in the summer and monsoon seasons. As a result large variations in meltwater discharge are common (Kumar et al. 2002). Monsoon precipitation, which falls as rain under warm air temperature conditions, also accelerates the melting of new snow cover (Hasnain and Thayyen 1999).

Besides the glacial basins of Himalayan Rivers, there are non-glacial (Rain fed) systems, which contribute significant amount of water to these river systems. In fact, most of the Himalayan basins are combination of the both the systems. Glacier basins have high energy and characteristic landforms with high elevated rocky terrain and presence of snow and ice (Hewitt 1972; Barsch and Caine 1984), whereas the non-glacial watersheds have lower elevation with gentle slopes, medium to good soil depth and intensive biotic activities. Runoff patterns of both systems are different. The amount of runoff coming from glacier is the product of glacierized area and ablation rate.

Glaciers act as natural reservoirs storing water in a frozen state and releasing the most runoff during the summer when all other sources of water are at a minimum (Stenborg 1970; Fountain and Tangborn 1985). This seasonal variation characteristic mitigates low flow intervals and makes glacier runoff a valuable water resource for hydropower (Fountain and Tangborn 1985). Glacier runoff may have less influence on the annual average flow from a large basin, as the total runoff over a period of several years is determined largely by annual precipitation. Flow from non-glacier system peaks in July and August. Unlike non-glacier runoff, glacier runoff correlates better with temperature than precipitation, due to the dominant role of glacier melt compared to precipitation in summer runoff from glacierized basins. This is also the reason for the strong diurnal nature of glacier runoff. Non-glacial systems form the important resources for the local population, whereas glacial meltwater flow has more regional implications. Very little work is done to understand and compare the hydrological behaviour of these two systems which can provide insight into these systems for sustainable water management. The present paper makes such an attempt in two small watersheds of Uttarakhand Himalaya.

2 Study Area and Methodology

The glacial basin of Gangotri Glacier is studied, which is one of the largest glaciers in Himalaya located in the Uttarkashi district of Uttarakhand state. Its watershed extends from 79° 17′ 18″ to 79° 4′ 55″ E and 30° 43′ 10″ to 30° 55′ 50″ N. It originates at a height of about 7,100 m above mean sea level (amsl) and, its terminus is at 3,922 m above mean sea level (amsl). Numerous small-sized glaciers join the main Gangotri glacier from all sides to form the Gangotri group of glaciers (after Sharma and Owen 1996; Fig. 1a). The main glacier, together with its tributaries, is a compound valley glacier. The total spread of the ice volume is about 39.18 km^3 (Kaul 1999) and the total glacierised area (75 % of total basin area) is about 258.56 km^2 (Naithani et al. 2001). The main glacier drains towards the northwest from the Baghirathi group of peaks. The basin is largely devoid of vegetation and human population.

As a non-glacial basin, the study is conducted in the northern part of the Kosi basin (upper Kosi watershed between 29° 30′ and 29° 55′ N Latitudes and 79° 30′ and 79° 45′ E Longitudes covering 480.15 km^2 area) spreading over the Lesser Himalayan domain and administratively within district Almora, Uttarakhand state.

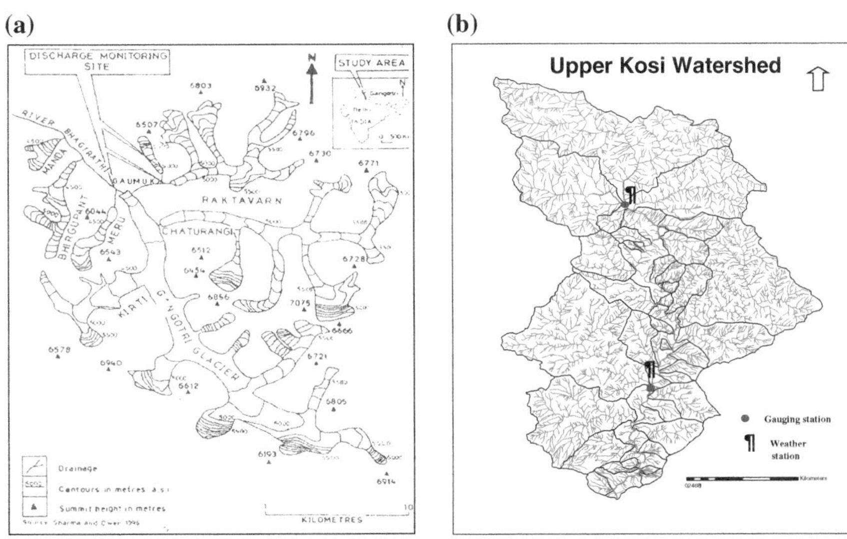

Fig. 1 **a** Glacial watershed. **b** Non-glacial watershed (Upper Kosi)

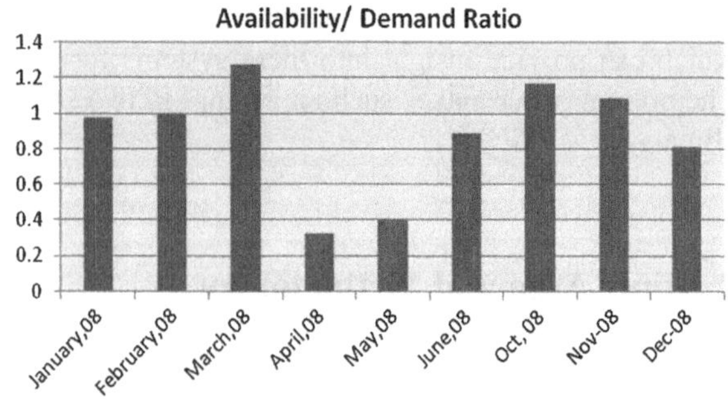

Fig. 2 Water availability demand ratio of Upper Kosi basin

The absolute relief of the catchment ranges between 1,080 m and 2,720 m from the mean sea level (Fig. 1b).

The basin is highly populated and water availability/demand ratio is below 1 in most of the dry months (Fig. 2) indicating severe water stress in non-monsoon months. A gauging station was established for the glacier discharge measurements on the Bhagirathi River, at 500 m downstream of the terminus of Gangotri glacial and for non glacial discharge measurement gauging station was established at Kosi River near Kosi Bridge. Automatic stage level recorder has been installed at gauging station for stage measurement. Discharge measurements were made using the area-velocity method by establishing a rating curve for both watersheds (Water Measurement Manual 1997).

3 Results

Flow pattern of glacial and non-glacial basins are discussed separately in the following sections in terms of daily discharge variations, flow duration curves and ratio of maximum and minimum discharge.

3.1 Rainfall and Meltwater Discharge from Glacial Basin

Daily rainfall was recorded for eleven years (1999–2009). Maximum total rainfall recorded in 2000 with value 306.14 mm and minimum total rainfall was recorded in 2004 having value 106.30 mm. Monthly rainfall for five years indicates that most of the rainfall has occurred in the month of June, July and August corresponding with the summer monsoon season. Average total rainfall over the entire monitoring period was 204.30 mm, which is very low in comparison to other valleys. Overall the Gangotri valleys received scanty rainfall due to its lee side location which forms a rain shadow zone. However, some events of high intensity rainfall were noticed in this area (i.e., 60.55 mm/day in June 2000) (Fig. 3). Such events were responsible for sudden increase in discharge and high concentration of suspended sediment into the river.

Himalayan Rivers along with glaciated catchments have regional importance, and the water from the glacier melt of sustains stream flow in such rivers through

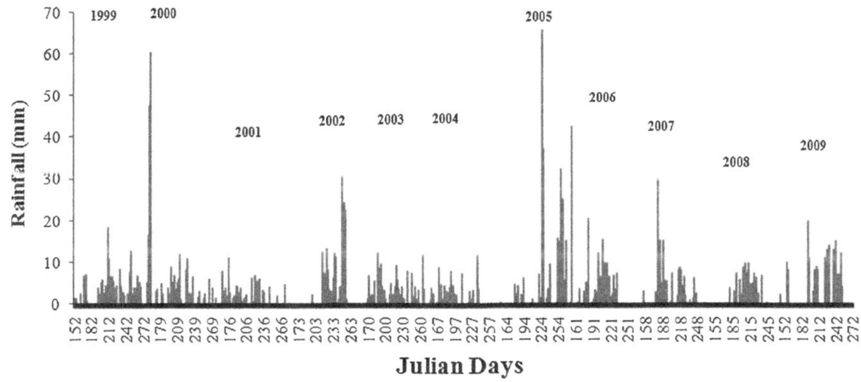

Fig. 3 Daily rainfall in Gomukh

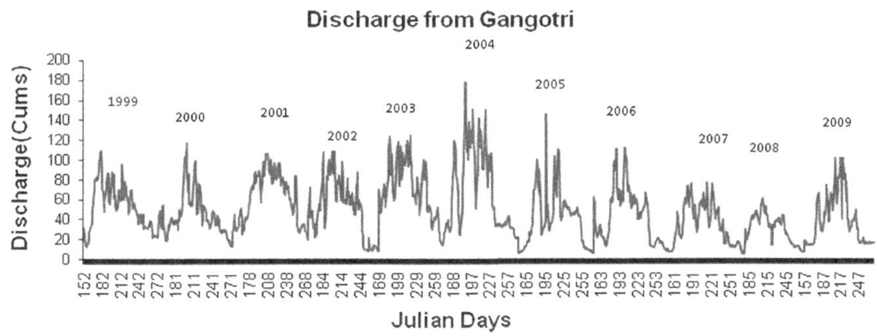

Fig. 4 Discharge pattern of Upper Bhagirathi basin (1 km downstream of Goumukh)

the dry season. Considering the importance of this "Water Tower" of the subcontinent, the study of its water resources becomes relevant in the context of changing climate and socio-physical conditions. Melt water discharge of Gangotri glacier in eleven consecutive years (1999–2009) showed variations on a seasonal scale in terms of Coefficient of Variation (CV) which is ranges from 0.40 to 0.70 during ablation season (June–September) when active melting takes place (Fig. 4). The distribution of mean monthly discharge in the channel showed that it started increasing from the month of June till mid-August and after that decreasing trend was followed in September, maximum and minimum discharge values behaved differently in the beginning of the season. Large variations were noticed in different months. Events related to high flow mostly occurred in the months of July and August confirming that these events were caused by opening up of drainage network and excessive melting of glacier. Analysis of hydrographs for different years indicates that sudden peaks in the stream flow are observed at the monitoring site. Such peaks in the hydrographs are often generated by outburst of water bodies formed in different part of glacier valley caused by rising temperature in late summer. Sudden high rainfall events can also cause rise in stream flow in the glacier basin.

In case of large glacier systems like Gangotri glacier, several discharge peaks can also be generated due to variation in melting rate of various tributary glaciers experiencing different temperature patterns. Though the rainfall is very low in valley, but distribution of rainfall also controls the runoff behaviour from the basin to some extent. Relatively higher runoff in July and August is due to influence of higher temperature in those months in the respective years. In the beginning of season, early rise in the hydrograph in every ablation season (1999–2009) was due to opening of ice jam conduits and their first flushing to downstream in the month of June. Flow duration curves constructed for ablation season (JJAS) indicate gentle slope (Fig. 5).

This indicates influence of glacier storage and release of water from different zones of the glacier which sustain the flow in the stream. The variations in mean daily flow in different months are lower. However, there are significant variations in the mean values of different months. The ratio of minimum and maximum discharge in different months of ablation season indicates high variations in June (Fig. 6). This is mainly

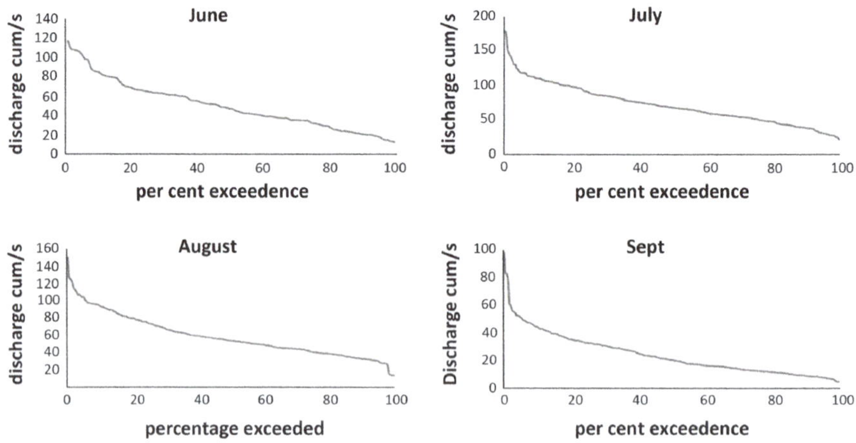

Fig. 5 FDC of glacial basin on Gangotri glacier

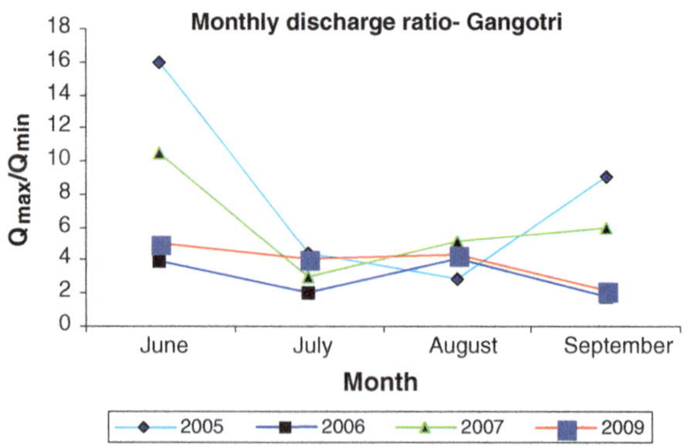

Fig. 6 Ratio of max and min discharge of gangotri basin

due to sudden peaking of discharge possibly due to opening of ice jammed conduits and flash floods in 2006 and 2007. However, the ratio is lower in subsequent monsoon months (July and August). Some increase in ratio in September is mainly due to low value of minimum flow at the end of the season. This is typical behaviour of glacial basins where flow is maintained by constant supply from stored frozen water.

3.2 Rainfall and River Discharge from Non-glacial Basin

The Kosi River is non-glacier river system which originates from forest of Pinath near Kausani (Almora district). The river flow is largely depends on rainfall. The mean rainfall of the basin is 1,400 mm and about 80 % of it is received in four monsoon months. Upper Kosi watershed is targeted by the local govt. agencies as water stressed considering the threat to water availability from the watershed, which is the only source for the area. Hourly rainfall-runoff relationships in monsoon months indicate that, runoff generated due to monsoon rainfall sustained for short duration (Fig. 7). There is average time lag of 1 day between rainfall and runoff for the watershed.

The flow duration curves of Kosi for different monsoon months describe the changing flow pattern (Fig. 8). A large part of the year falls under dry weather and low flow conditions and the flow generated by the rain recedes very fast. Only about 10 % time in the year the high flow conditions are observed. For almost 60 % of the time the river flow is sustained by the base flow.

The rate of increase and decrease of discharge with time (dry and wet season) depends upon type of source, storage characteristics of the aquifer and catchment area. In 2008, the Qmax /Qmin ratio was the highest in June (240.12) and it was maximum in May having a value of 302.06 in 2009. In 2010, this ratio was maximum

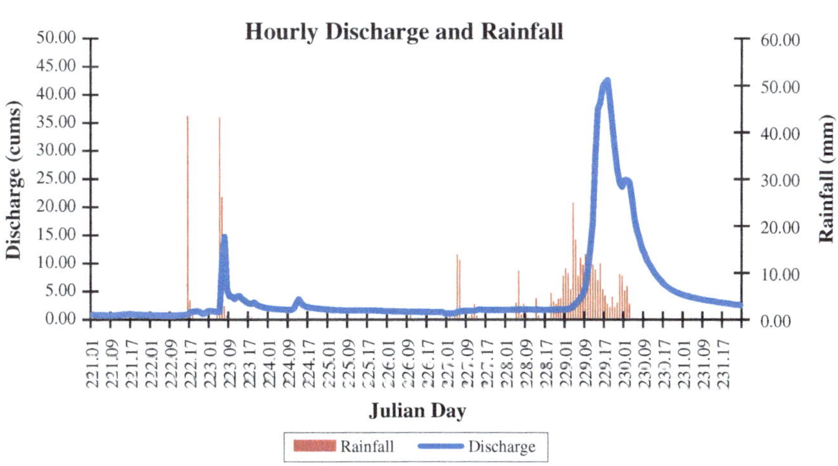

Fig. 7 Hourly discharge and rainfall in monsoon season at Upper Kosi catchment

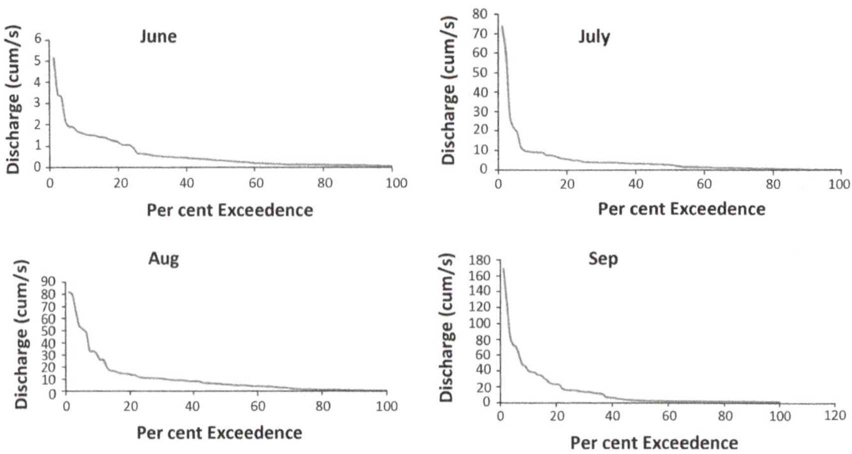

Fig. 8 Flow duration curve for Upper Kosi basin

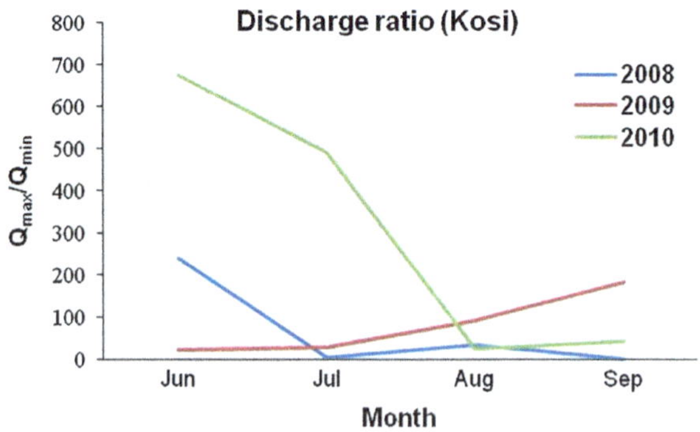

Fig. 9 Ratio of max and min discharge of upper koshi basin

in June (Fig. 9). There is a growing trend in Q_{max}/Q_{min} ratio and low discharge days are increasing, however, longer data may be required to establish the trend. The influence of rainfall is clearly visible in variation of discharge ratios in different years.

3.3 Hydrological Response of Non-glacial Basin

The non-glacial basins are typically modified by the human interventions and facing severe water and resource scarcity. The hydrological responses of different altitudinal zones are also not same except for the decline in stream and spring discharges (Table 1). Lower valleys are densely populated and intensive agriculture is practiced there. Whereas, in middle and uplands the agriculture is mostly rainfed (RF) and

Table 1 Hydrological characteristics of non-glacial basin

Lower transect (below 800 m)	Middle zone (800–1,600 m)	Upper zone (1,600–2,400 m)
Water discharge decreasing (up to 20 %)	Water discharge decreasing (up to 30 %)	Water discharge decreasing (up to 35 %)
Catchment area (15–20 ha) in RF Agri. and multi village	Catchment area (5–12 ha) in civil forest, Agri. and multi village	Catchment area (8–18 ha) in civil forest, one more village and RF
Dense mixed reserve forest	Open Pine forest	Dense mixed forest
Deforestation has limited role in source decline	Deforestation has limited role in source decline	Deforestation has major role in source decline
Flooding, land slide, gulley erosion and slumming in catchment area. Medium slope	Gulley erosion in catchment area with moderate to steep slope	Land slide and gulley erosion with steep slope
Limited grazing pressure on the catchment area	Heavy grazing pressure on the catchment area	Grazing pressure on the catchment area
Catchment area is enough to meet water demand. Large population depends on catchment	Catchment area only sufficient to meet demand at present, if conserved	Catchment area is large to meet demand. Catchment dependence is less

productivity is less. These areas show abandoned common and private lands experiencing heavy soil erosion and nutrient losses. Water retention capacity is reduced due to the action of erosion and deforestation. Most of these areas demand urgent action for rehabilitation and watershed management to sustain water regime.

4 Conclusions and Recommendations

The study highlights the specific features of the hydrologic behavior of a glacial and non-glacial basin. Glacial basins are characterized as high energy landforms with less biotic activities, whereas the non-glacial basins have gentle slopes and more intensive biotic activities. Hydrological responses of the studied basins confirm the role of these characteristics. Glacial basin generates large flow variations on seasonal basis, but the mean daily flow in different months is less variable. The role of frozen storage and difference in time of concentration is significant for maintaining the flow in compound glacier like Gangotri. In case of non-glacial systems, the flow is more precipitation dependent and catchment characteristics play an important role in time of concentration. Headwaters of such rivers have time lag of only few days and flow is not sustained after rains as reflected by the high discharge ratios in case of Kosi. This has serious implications for water management in such water stressed systems. Based on the study some management and adaptation options for such basins are recommended.

1. Water management and conservation during extreme events is a top priority of the glacial watersheds. In the absence of proper planning high intensity rainfall

distributed over a small geographical area (cloud bursts), outburst of glacial (GLOF) and landslide dammed lakes can give rise to floods and associated loss of lives and property in the downstream regions. There is a need of appropriate technologies to harness the non-consumptive potential of water resources in this region. There is still a scope to tap the huge water potential in the rivers which has not been adequately utilized so far for irrigation and power generation, adventure tourism, fish cultivation and other industrial uses.
2. Optimisation of water allocation is needed in non-glacial systems for its efficient utilisation for population, livestock, production of non-agricultural and industrial items, production of energy, improvement of the quality of life and preserving the ecology of the region. One of the major challenges on the way to sustainable management of water is providing equitable access to water, both in terms of quality and quantity.
3. Information sharing mechanism for water management is urgently needed in institutional level. There is lack of well-developed meteorological and river gauging network for improving the knowledge on hydrology, rainfall and sediment transport, etc. throughout the region. A continuous monitoring of hydro-meteorological data will help the planners for their planning and making projections for water management for future.
4. Glacier retreat is a growing concern throughout the Himalayan region. Spatio-temporal monitoring of glacier dynamics and their melting rate is required to improve the designs and operational efficiency of all proposed hydropower and irrigation projects. This knowledge will also help for formulating efficient strategies for conservation, mitigation and adaptations in the downstream.

References

Barsch D, Caine N (1984) The nature of mountain geomorphology. Mt Res Dev 4:287–298

Bruijinzeel A, Bremmer NN (1989) Highland and low land interaction in the Ganges–Brahmaputra river basin: a review of published literature. Occasional paper no. 11, International Center for Integrated Mountain Development (ICIMOD), Kathmandu, Nepal

Collins DN, Hasnain SI (1994) Runoff and sediment transport from the glacierized basins at the Himalayan scale. In: Osterkamp WR (ed) Effects of scale on interpretation and management of sediment and water quality. Proceeding, boulder bymposium, international association of hydrological sciences, pp 17–25

Fountain AG, Tangborn WV (1985) The effect of glaciers on streamflow variations. Water Resour Res 21:579–586

Hasnain SI, Thayyen RJ (1999) Discharge and suspended sediment concentration of meltwaters, draining from the Dokriani Glacier, Garhwal Himalaya. J Hydrol 218:191–198

Hewitt K (1972) The mountain environment and geomorphic processes. In: Slaymaker HO, Mcpherson HJ (eds) Mountain geomorphology: geomorphological processes in the Candian Cordillera, B. C. Geomorphological Series No. 14, Tantalus Research, Vancouver

Kaul MK (1999) Inventory of the Himalayan Glaciers. Geological Survey of India (GSI). Special Publication No. 34, p 165

Kumar K, Miral MS, Joshi V, Panda YS (2002) Discharge and suspended sediment in the meltwater of Gangotri glacier in Garhwal Himalaya. Hydro Sci J 47(4):611–619

Naithani AK, Nainwal HC, Sati KK, Prasad C (2001) Geomorphological evidence of retreat of the Gangotri Glacier and its characteristics. Curr Sci 80(1):87–88

Raina VK, Srivastava D (2008) Glacier atlas of india. Geological Society of India. Pragati Graphics, Bangalore

Sharma MC, Owen LA (1996) Quaternary glacial history of NW Garhwal, Central Himalaya. Quat Sci Rev 5:335–365

Stenborg T (1970) Delay of runoff from a glacier basin. Geografi ska Annaler 52 A, pp 1–30

Valdiya KS (1998) Dynamic of Himalaya. Universities press, Hyedrabad, p 178

Water Measurement Manual (1997) Edition 3. Scientific Publishers, Jodhpur, p 400

Variable Response of Glaciers to Climate Change in Uttarakhand Himalaya, India

Dwarika P. Dobhal and Bhanu Pratap

Abstract The glaciers are fragile and dynamic in nature and influence the climate system (e.g. albedo feedback) and as well as key indicator of climate change. The reduction in mass, volume, area and length of glaciers are considered as clear signals of a warmer climate. Uttarakhand Himalaya contains 968 glaciers out of the total 9,575 glaciers in Indian part of the Himalaya, covering an area of 2,888.37 km^2 with 213.74 km^3 of ice volume lies between the altitudes 6,600 and 3,860 m with different dimensions. The observations made during the end of nineteenth century over the Uttarakhand Himalayan glaciers indicate that there is continuous retreat of glaciers but rate of retreat are different to different glaciers. In this study, the results of a detailed mapping campaign and ground-based measurements of terminus retreat, area vacated and mass/volume change has been carried out on few glaciers for the period between 1962 and 2010. The study shows continuous negative mass balance on Tipra, Dunagiri, Dokriani and Chorabari glaciers during last three decades. In general, Uttarakhand Himalayan glaciers are under substantial thinning (Mass loss) and reduction of length and area in the present climate conditions.

Keywords Glacier · Retreat · Mass balance · Climate change · Uttarakhand Himalaya

1 Introduction

Several future scenarios have been predicted for the climate change and future of Himalayan glaciers speculating the trends and consequences (Bajracharya et al. 2008). Change in temperature and snowfall pattern have been observed in the

D.P. Dobhal (✉) · B. Pratap
Centre for Glaciology, Wadia Institute of Himalayan Geology,
Dehradun, Uttarakhand, India
e-mail: dpdobhal@wihg.res.in

R. Joshi et al. (eds.), *Dynamics of Climate Change and Water Resources of Northwestern Himalaya*, Society of Earth Scientists Series,
DOI 10.1007/978-3-319-13743-8_12

Himalaya during last century. For instance, significant rise of 1.6 °C temperature between 1901 and 2002 has been reported in the Northwestern Himalaya (Bhutiyani et al. 2007). The seasonal mean, maximum and minimum temperatures of winter months have also been increased over the western Himalaya from 1985 to 2008 and pattern of snowfall has been decreased in winter during similar period (Shekhar et al. 2010). Variation in climatic parameters (e.g. temperature and precipitation) directly affects the glacier mass balance and its dynamic properties. Therefore, glaciers are considered to be as a sensitive indicator of climate (Dobhal et al. 2004) and good indicator of energy balance (Oerlemans 1989).

A study by Mayewski and Jeschke (1979) indicates that generally Himalayan glaciers have been receding since 1850. Several studies have also reported the increase in glacier recession rate and continuous negative mass balance in the last decades especially in the central Himalaya (Dobhal et al. 2008; Kulkarni et al. 2007; Bolch et al. 2012; Bhambri et al. 2011). Recently, a study has reported significant ice mass loss for smaller glaciers concurrently increased number of glaciers due to fragmentation of the tributary glaciers in the Bhagirathi and Alaknanda basins (Bhambri et al. 2011). In addition, Kulkarni et al. (2007) and Dobhal et al. (2013b) have described the role of debris thickness in the lower region of the glaciers that can reduce the retreat of glacier but can leads to fragmentation of the snout. Due to glaciers retreat, favorable conditions for formation of lakes have increased and generally form behind the newly exposed terminal moraine.

The rapid accumulation of water in these lakes can lead to a sudden breach of the moraine dam. The resultant rapid discharge of huge amounts of water and debris is known as a glacial lake outburst flood (GLOF) and the results can be catastrophic to the downstream (Richardson and Reynolds 2000; Dobhal et al. 2013a). One recent glacier inventory in Alaknanda and Bhagirathi suggest that ~25 % of total glacial area is covered by debris in their ablation zones (Bhambri et al. 2011). Thus, the regular monitoring of these glaciers of this region is important as water discharge from these glaciers and snow melt contributes to a significant amount to the overall river runoff of Ganga (Immerzeel et al. 2010) and provide noteworthy contribution in hydropower development in Uttarakhand state.

2 Study Area

The Indian Himalaya with various climatic zone consist 9,575 glaciers covering an area of ~40,000 km^2 with an ice volume of ~2,000 km^3 (Raina and Srivastava 2008). Himalayan glaciers in the Indian subcontinent are broadly divided into the three river basins namely the Indus, Ganga and Brahmaputra. The Indus basin has the largest number of glaciers (~7,997), whereas the Ganga basin including Brahmaputra contain about 1,578 glaciers (Raina and Srivastava 2008). Out of these, Uttarakhand Himalaya covers 968 glaciers covering 2,888 km^2 (Raina and Srivastava 2008). This show ~13.8 % glacier area of entire Indian Himalaya exists in Uttarakhand.

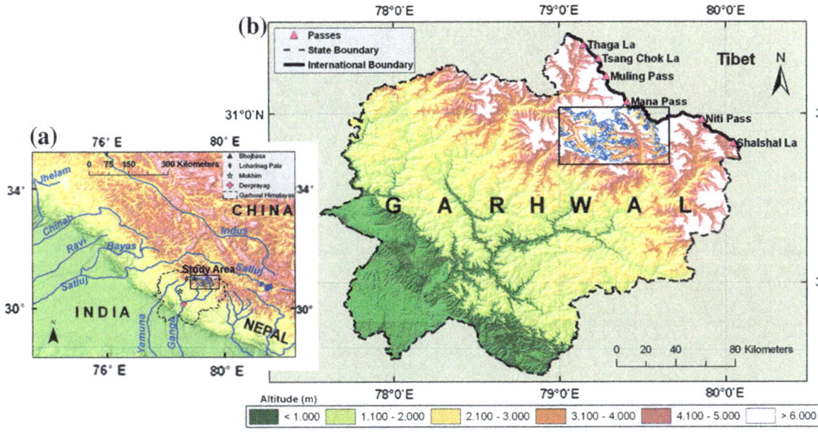

Fig. 1 Location of Uttarakhand Himalaya in India

The Uttarakhand Himalaya covers Yamuna, Bhagirathi, Alaknanda and Kali Ganga (Ghaghra) basins (Fig. 1). The Bhagirathi River is the main source stream of the Ganga River, which originates from the snout (Gaumukh; ~3,950 m a.s.l.) of Gangotri Glacier, the largest valley glacier (~30 km) in the Uttarakhand Himalaya. The headwater of the Alaknanda River originates from the snouts of Bhagirath Kharak and Satopanth glaciers. The Alaknanda basin has ~407 glaciers covering ~1,255 km^2, whereas the Bhagirathi basin contains ~238 glaciers covering ~759 km^2 (Sangewar and Shukla 2009). Uttarakhand Himalayan glaciers are fed by summer monsoon and winter snow regimes (Thayyen and Gergan 2010). However, maximum snowfall occurs from December to March, mostly due to western disturbances (Dobhal et al. 2008).

3 Glacier Variations in the Uttarakhand Himalaya

3.1 Length Changes

The Geological Survey of India (GSI) and other scientific Indian organizations have at their disposal almost 100 years of well-documented recession records of the lengths of selected Uttarakhand Himalayan glaciers, such as Gangotri (1842–2006), Meola (1912–2000) and Milam (1906–1997) glaciers. Gangotri and Dokriani glaciers have been surveyed extensively by several researchers working on mass balance, hydrological, geomorphological, isotopic and frontal recession. The cumulative length recession records of Uttarakhand Himalayan glaciers (Fig. 2) indicate that glacier retreat is irregular in extent and rate. Therefore, these records have to be used with caution owing to the different response time of the glaciers. For instance, in Bhagirathi basin, Dokriani Glacier retreated at the rate of 16.5 a^{-1} during 1962–1991, 17.8 a^{-1} from 1991 to 2000, and 15.75 m a^{-1} for the periods of 2000–2007 (Dobhal and Mehta 2008).

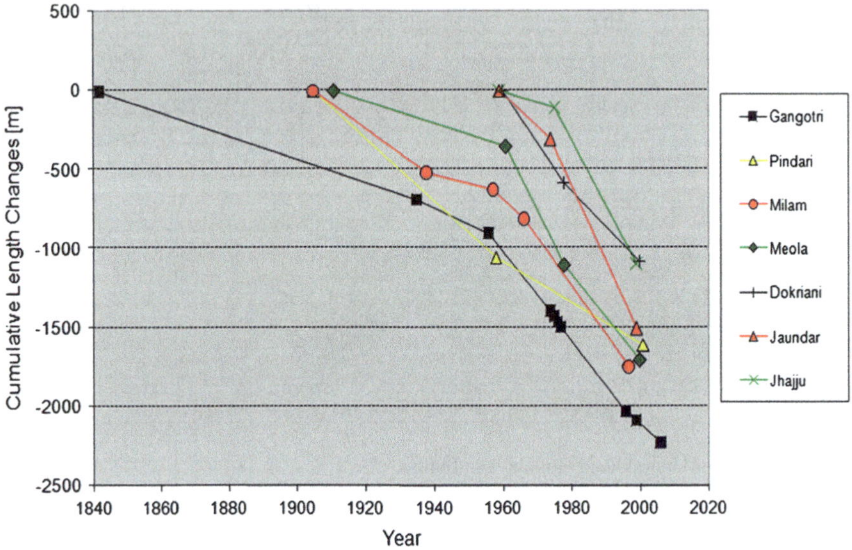

Fig. 2 Length records of selected Uttarakhand Himalayan glaciers (*Source* Bhambri and Bolch 2009 modified)

During the study period of 1996–2007 no advancement was found indicating that the glacier is receding continuously. Available fluctuation records of Indian glaciers suggest that commonly, south, southeast and southwest facing glaciers, such as Jaundhar, Jhajju and Tilku Glaciers in the Tons Valley, have receded rapidly at a rate of more than 34.18, 15.38 and 13.46 m a^{-1} respectively between 1962 and 2010 (Mehta et al. 2013). Measurement of snout positions of the Tipra and Rataban Glaciers from 2002 to 2008 indicates an enhanced annual retreat of 21.3 and 21.2 m a^{-1}, respectively (Mehta et al. 2011). These valley glaciers accumulate under the influence of the southwestern monsoon. The study suggest that glaciers longer than 15 km retreat at a rate of more than 20 m a^{-1}, except for Millam and Gangotri glaciers and furthermore, the recession of these glaciers has accelerated compared with the previous observation (Fig. 3).

However recession of Gangotri Glacier in last decade shows different behavior as compare to relatively small glaciers in Uttarakhand Himalaya. This glacier retreated 819 ± 14 m from 1965 to 2006 (Bhambri et al. 2012). On an average, Gangotri Glacier retreated at the rate of 5.9 ± 4.2 m/year from 1965 to 1968 and 26.9 ± 1.8 m/year from 1968 to 1980, and it retreated 21.0 ± 1.2 m a^{-1} between 1980 and 2001. The recession rate declined during 2001–2006 and it receded at a rate of 7.0 ± 4.0 m/year. From 2001 to 2006, the recession of Gangotri Glacier has declined compared to the previous observation during the study period. However, it does not imply that Gangotri Glacier recession has ceased as length changes show only the indirect and delayed response of a glacier to climate change, in contrast to glacier mass balance. The response time of the large debris-covered Gangotri Glacier is likely to be much longer than that of smaller

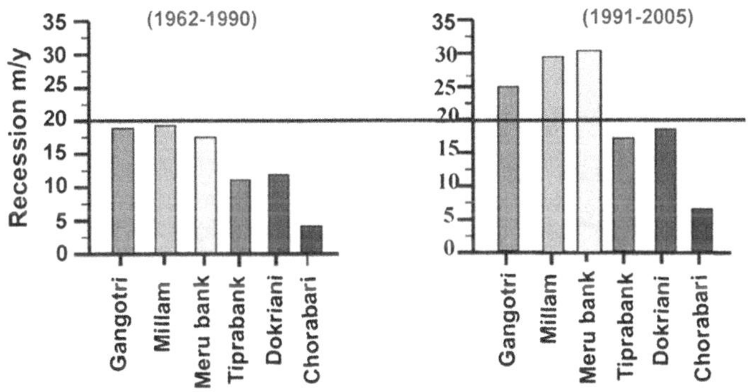

Fig. 3 Changes of Snout retreat rates of glaciers in Uttarakhand Himalaya during the last few decades

glaciers in the Uttarakhand region. Considering rate of recession of these glaciers, it has been observed that they are generally in state of recession, however rate of recession varied from 5 to 20 m/year. A study of long term monitoring glaciers in Uttarakhand Himalaya clearly show that the retreat rate has been enhance during the last few decades but there is no change in their trend (Fig. 3).

3.2 Glacier Area Changes

Glacier area change can be analyzed in terms of frontal area and entire area loss. In the Alaknanda basin, the ice-covered area of Satopanth Glacier diminished by 0.314 m^2 (1.5 %) near the snout from 1962 to 2006, whereas Bhagirathi Kharak Glacier lost an area of 0.129 m^2 (0.4 %) during a similar time period (Nainwal et al. 2008). These two glaciers are situated in the same basin and likely experienced similar climatic conditions. However, both glaciers are reported to be retreating at different recession rates. Bhagirathi Kharak and Satopanth glaciers retreated 7.4 and 22.8 m a^{-1} during 1962–2006, respectively. This might be due to uneven distribution of tributary glaciers, active cirques, drainage density and distributions of supra-glacial debris cover (Nainwal et al. 2008). One recent study reveals that Tipra and Rataban glaciers vacated their frontal area by 0.084 and 0.028 km^2 respectively between 1962 and 2002 (Mehta et al. 2011).

Bhambri et al. (2011) recently reported glacier area decreased from 600 ± 15.6 km^2 (1968) to 572.5 ± 18.0 km^2 (2006), a loss of 4.6 ± 2.8 % in Uttarakhand Himalaya. In this study, glaciers in Saraswati/Alaknanda basin and upper Bhagirathi basin lost 18.4 ± 9.0 km^2 (5.7 ± 2.7 %) and 9.0 ± 7.7 km^2 (3.3 ± 2.8 %) respectively, from 1968 to 2006 (Fig. 4). The number of glaciers in this study increased from 82 in 1968 to 88 in 2006 due to fragmentation of glaciers. In addition, smaller glaciers (<1 km^2) lost 19.4 ± 2.5 % (0.51 ± 0.07 % a^{-1}) of their ice, significantly more than for larger glaciers

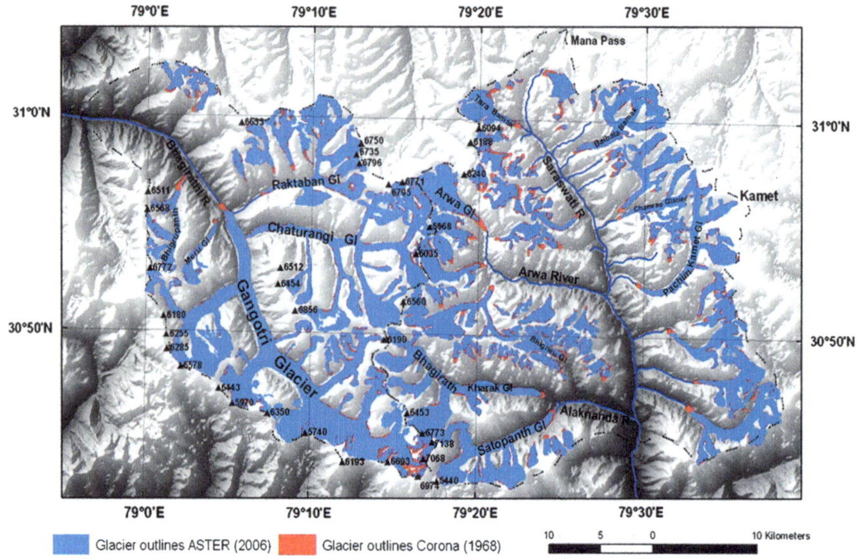

Fig. 4 Glacier change in the Uttarakhand Himalaya from 1968 to 2006 (*red* and *blue* glacier outlines derived from Corona and ASTER images respectively

(>50 km^2) which lost 2.8 ± 2.7 % (0.074 ± 0.071 % a^{-1}). From 1968 to 2006, the debris-covered glacier area increased by 17.8 ± 3.1 % (0.46 ± 0.08 % a^{-1}) in Saraswati/Alaknanda basin and 11.8 ± 3.0 % (0.31 ± 0.08 % a^{-1}) in the upper Bhagirathi basin. Individual study on Dokriani Glacier shows that this glacier lost about 9.5 % area during 1962–2007 (Dobhal and Mehta 2010).

3.3 Mass Balance Studies

Glacier mass balance is the in situ measurements of accumulation and ablation of entire glacier during a balance year that provide immediate indication of storage system (Paterson 1994). Changes in glacier mass over years reflect the behavior of the glaciers. When a large quantity of snow accumulates in winter survives and exceeds ablation, the glacier shows mass gain. Similarly, if melting is more pronounced than the accumulation, the balance shows negative response and subsequently glacier losses it's mass (Kasser et al. 2003; Dobhal et al. 2008; Cogley et al. 2011). Ideally, glacier mass balance is monitored continuously for many hydrological years to derive volume loss or gain to understand climatic fluctuations. Thus, cumulative mass balance trends of several years indicate regional climatic variability (Hodge et al. 1998; Dyurgerov and Meier 2005). In Uttarakhand Himalaya, so far only four glaciers have been considered for mass balance estimation by field based glaciological methods: Tipra Bank (1982–1989), Dunagiri

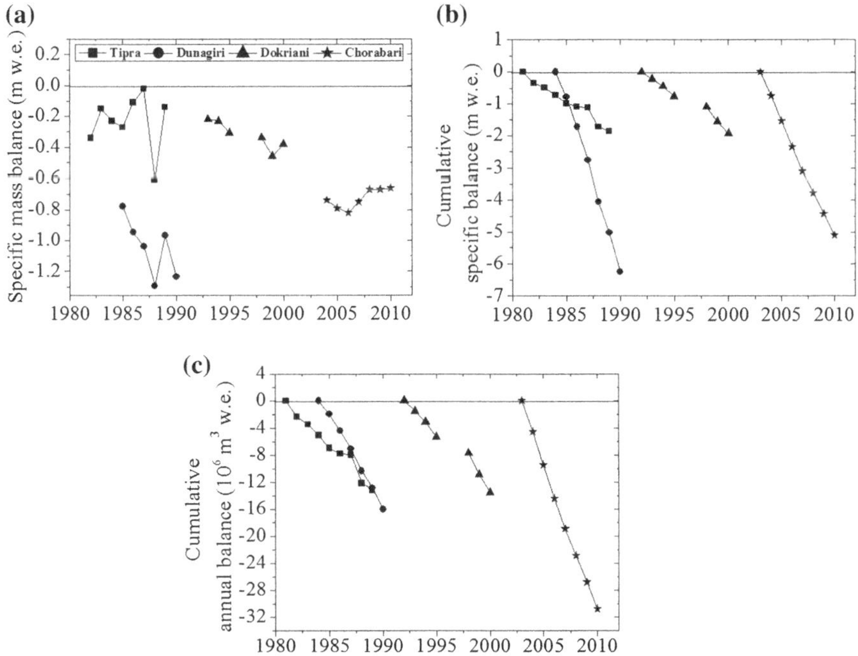

Fig. 5 **a** Annual specific mass balance, **b** cumulative specific mass balance and **c** cumulative net glacier-wide mass balance of 4 Uttarakhand Himalayan glaciers

(1985–1990), Dokriani (1993–2000) and Chorabari (2004–2010). Mass balance measurements of Tipra Bank (7 km^2) and Dunagiri (2.5 km^2) glaciers shows average thinning of −0.23 and −1.04 m a^{-1} w.e. during 1980s (Fig. 5a, b). Mass balance analyses in 1990s on Dokriani Glacier indicate continuous negative annual mass balances. From 1992/1993 to 1998/1999 the trend was increasingly negative; −0.22 to −0.44 m w.e. respectively. The cumulated mass balance reached −1.94 m w.e. from 1993 to 2000 which shows −0.32 m a^{-1} w.e. ice mass loss of the glacier. The calculated ice volume of Dokriani Glacier was 385.11×10^6 m^3 w.e. in 1962 and 315.0×10^6 m^3 w.e. in 1995 suggests a total reduction of -70.11×10^6 m^3 w.e. ice mass in 33 years.

The annual mass balance from 1995 to 2000 shows continuous negative mass balances, which tend to compute total loss of -12.0×10^6 m^3 w.e. in 5 years. Therefore, total volume change is -82.11×10^6 m^3 w.e. for Dokriani Glacier since 1962–2000. In the first decade of 21st century (2000–2010) annual mass balances was measured for Chorabari Glacier during 2004–2010 (Fig. 5a). Over these 7 years glacier showed mean annual balance of −0.73 m w.e. a^{-1}. This leads to total cumulative mass balance of −5.1 m w.e. from 2004 to 2010. The annual specific mass balance investigated during the study period is with the maximum deficit (−0.82 m w.e.) in 2005/2006 and minimum deficit of (−0.65 m w.e.) in 2009/2010. The thinning is the main component of glacier shrinkage. Therefore, during the periods of 7 year the

Chorabari Glacier has reduced its ice thickness by 6 m. Field based studies on the two glaciers (Dokriani and Chorabari) in Bhagirathi and Mandakini basins provide great insight into the heavy rainfall events during monsoon periods of 2011 and 2013 in short time span, which caused large amount of sediment transport as well as catastrophic events of debris flow and moraine dammed Lake Outburst near the glaciers snout.

4 Conclusions

Glacier variability depends on several environmental conditions such as minimum and maximum temperature, solar radiation, amount of solid precipitation and debris cover load, as well as altitude and orientation of the glaciers, which influence the mass balance so as the length change of the glaciers. Out of 968 glaciers in the Uttarakhand Himalaya only few glaciers were monitored for the long term mass balance and snout recession. Overall, studies of glacier length and area change in Uttarakhand Himalaya indicate that larger glacier has less variation rather than small glaciers. Therefore, it can be concluded that glaciers length change is largely dependent on the shape, size and thickness of the glacier. Glacier mass balance study was conducted only for four glaciers in the Uttarakhand Himalaya during 1982–2010. The result shows continuous negative mass balance trend along with variability in net annual mass balance. Maximum cumulative ice mass loss is observed in Chorabari Glacier i.e. -30.77×10^6 m^3 w.e. in 7 years (2004–2010) with -4.4×10^6 m^3 w.e. a^{-1} resulting -0.73 m w.e. y a^{-1} of specific mass balance (Fig. 5a–c). Differences in mass balance observations of individual glacier may be due to variability in the orientation and topographic regime of the region besides the climate.

As a result of negative mass balance and vacating glacierised area, field based studies on the two glaciers (Dokriani and Chorabari) during 2011 and 2013 emphasized catastrophic events of debris flow sediment transport as well as moraine dammed Lake Outburst near the glaciers snout. In general, glaciers in Uttarakhand Himalayan are under substantial ice melting in the present climate conditions. However, existing data exhibit short time series biases to make out a complete scenario of climate change impact on Himalayan glaciers. Thus, there is a need to design for long term glaciers monitoring network to understand the variability of glaciers mass changes over Himalaya.

Acknowledgments The authors are grateful to Prof. A.K. Gupta, Director, Wadia Institute of Himalayan Geology (WIHG), Dehra Dun, India for providing facilities and ample support to carry out this work. We are also thankful to Department of Science and Technology, Government of India for funding.

References

Bajracharya SR, Mool PK, Shrestha BR (2008) Global climate change and melting of Himalayan glaciers. In: Ranade PS (ed) Melting glaciers and rising sea levels: impacts and implications. ICFAI University Press, Hyderabad, pp 28–46

Bhambri R, Bolch T (2009) Glacier mapping: a review with special reference to the Indian Himalayas. Prog Phys Geogr 33(5):672–704

Bhambri R, Bolch T, Chaujar RK (2012) Frontal recession of Gangotri glacier, Uttarakhand Himalayas, from 1965–2006, measured through high resolution remote sensing data. Curr Sci 102:489–494

Bhambri R, Bolch T, Chaujar RK, Kulshreshta SC (2011) Glacier changes in the Uttarakhand Himalaya, India 1968–2006 based on remote sensing. J Glaciol 57:543–556

Bhutiyani MR, Kale VS, Pawar NJ (2007) Long term trends in maximum, minimum and mean annual temperatures across the northwestern Himalaya during the twentieth century. Clim Change 85:159–177

Bolch T, Kulkarni A, Kaab A, Huggel C, Paul F, Cogley G, Frey H, Kargel JS, Fujita K, Scheel M, Bajracharya S, Stoffel M (2012) The state and fate of Himalayan glaciers. Science 336:310–314

Cogley JG, Hock R, Rasmussen LA, Arendt AA, Bauder A, Braithwaite RJ, Jansson P, Kaser G, Möller M, Nicholson L, Zemp M (2011) Glossary of glacier mass balance and related terms, IHP-VII technical documents in hydrology No. 86, IACS contribution No. 2, UNESCO-IHP, Paris, p 114

Dobhal DP, Mehta M (2008) Snout fluctuation of Dokriani Glacier (1962–2007) vis-à-vis the impact of climate change. Himalayan Geol 29(3):23–25

Dobhal DP, Mehta M (2010) Surface morphology, elevation changes and terminus retreat of Dokriani glacier, Uttarakhand Himalaya: implication for climate change. Himalayan Geol 31(1):71–78

Dobhal DP, Gergan JT, Thayyen RJ (2008) Mass balance studies of the Dokriani glacier from 1992 to 2000, Uttarakhand Himalaya, India. Bull Glaciol Res 25:9–17

Dobhal DP, Gupta AK, Mehta M, Khandelwal DD (2013a) Kedarnath disaster: facts and plausible causes. Curr Sci 105:171–174

Dobhal DP, Mehta M, Srivastava D (2013b) Influence of debris cover on terminus retreat and volume change of Chorabari glacier, Garhwal region, central Himalaya, India. J Glaciol 59(217):961–971

Dobhal DP, Gergan JT, Thayyen RJ (2004) Recession and morphometrical change of Dokriani Glacier (1962–1995), Uttarakhand Himalaya, India. Curr Sci 86(5):692–696

Dyurgerov MD, Meier MF (2005) Glaciers and changing earth system: a 2004 snapshot. Institute of Arctic and Alpine Research, University of Colorado, Boulder, occasional paper No. 58, p 117

Hodge SM, Trabant DC, Krimmel RM, Heinrichs TA, March RS, Josberger EG (1998) Climate variations and changes in mass of three glaciers in western North America. J Clim 11(9):2161–2179

Immerzeel WW, Van Beek LPH, Bierkens MFP (2010) Climate change will affect the Asian water towers. Science 328:1382–1385

Kaser G, Fountain A, Jansson P (2003) A manual for monitoring the mass balance of mountain glaciers. IHP-VI technical documents in hydrology No. 59, UNESCO-IHP, Paris, p 106

Kulkarni AV, Bahuguna IM, Rathore BP, Singh SK, Randhawa SS, Sood RK, Dhar S (2007) Glacial retreat in Himalaya using Indian remote sensing satellite data. Curr Sci 92:69–74

Mayewski PA, Jeschke PA (1979) Himalayan and Trans Himalayan glacier fluctuations since AD 1812. Arct Antarct Alp Res 11(3):267–287

Mehta M, Dobhal DP, Bisht MPS (2011) Change of Tipra glacier in the Uttarakhand Himalaya, India, between 1962 and 2008. Prog Phys Geogr 30:1–18

Mehta M, Dobhal DP, Pratap B, Verma A, Kumar A, Srivastava D (2013) Glacier changes in upper tons river basin, Uttarakhand Himalaya, Uttarakhand, India. J Geomorphol 57(2):225–244

Nainwal HC, Negi BDS, Chaudhary M, Sajwan KS, Gaurav A (2008) Temporal changes in rate of recession: evidences from Satopanth and Bhagirath Kharak glaciers, Uttarakhand, using total station survey. Curr Sci 94:653–660

Oerlemans J (1989) A projection of future sea level. Clim Change 15:151–174

Paterson WSB (1994) The physics of glaciers, 3rd edn. Pergamon Press, Oxford

Raina VK, Srivastava D (2008) Glacier atlas of India. Geological Society of India, Bangalore 316

Richardson SD, Reynolds JM (2000) An overview of glacial hazards in the Himalayas. Quatern Int 65(66):31–47

Sangewar CV, Shukla SP (2009) Inventory of the Himalayan glaciers. Geological survey of India. Special publication No. 34

Shekhar MS, Chand H, Kumar S, Srinivasan K, Ganju A (2010) Climate change studies in the western Himalayas. Ann Glaciol 51:105–112

Thayyen RJ, Gergan JT (2010) Role of glaciers in watershed hydrology: a preliminary study of a Himalayan catchment. Cryosphere 4:115–128

Declining Changes in Spring Hydrology of Non-glacial River Basins in Himalaya: A Case Study of Dabka Catchment

Charu C. Pant and Pradeep K. Rawat

Abstract Natural springs are the key source of water for the stream flows within non-glacial river basins and constitute the main source of drinking water and irrigation. Therefore, assessment and inventory on spring hydrology are essential for formulation of successful sustainable development plans in the region. Consequently the main objective of the study was to assess the geohydrological processes of the springs using GIS technology. The Dabka catchment constitutes a part of the Kosi Basin in the Lesser Himalaya, India in district Nainital has been selected for the case illustration. The results suggested that most of the perennial springs exist along the thrust/fault planes and fluvial deposit areas and the most of the non-perennial springs exist along the fracture/joints and shear zones. The water yield of these springs varies greatly but maximum rate was monitored for thrust and fault controlled springs. The changing pattern of the spring hydrology suggested drying up of some springs and the reduced discharge of others due to accelerated land use changes by anthropogenic and climate change factors during last 2 decades (1991–2011). Consequently, there is a decline in the flow of the main Dabka River and its streams by 35 %.

Keywords Spring hydrology · Non-glacial river basins · Geohydrological processes · Dabka catchment · Himalaya

1 Introduction

Himalaya has rich geo-diversity and it plays a significant role in driving hydrological processes of springs through different types of lithology, structural lineaments, geomorphological layout, tectonic landforms, slope categories, drainage

C.C. Pant (✉)
Department of Geology, Kumaun University, Nainital, Utarakhand, India
e-mail: ccpgeol@yahoo.com

P.K. Rawat
Department of Geography, Kumaun University, Nainital, Utarakhand, India

R. Joshi et al. (eds.), *Dynamics of Climate Change and Water Resources of Northwestern Himalaya*, Society of Earth Scientists Series,
DOI 10.1007/978-3-319-13743-8_13

pattern, soils and land use pattern etc. (Rawat 2012). Degraded geo-diversity increases evapotranspiration, overland flow and groundwater inputs into streams (Bartarya 1991; Davies-Colley and Smith 2001; Rawat 1987; Sutherland et al. 2002). Investigations have revealed that increases in a watershed's proportion of impermeable surface such as hardened soils, surficial bedrock, or human-made structures such as pavement or buildings may considerably impact spring and stream hydrology and produce both high magnitudes and early peaks in storm hydrographs (Dunn and Lilly 2001; Singh and Bengtsson 2004; Nearing et al. 2005; Randhir 2005). These alterations in hydrology can have dramatic effects on ecological processes within stream ecosystems. Anthropogenic activities are continuously disturbing the natural system of the Himalayan environment, impacts of which can be seen in the hydrological behavior of streams and springs (Bisht and Tiwari 1996; Haigh et al. 1989; Hundecha and Bárdossy 2004; Marshall and Randhir 2008).

Fluctuation of water discharge in different seasons and increasing amount of sediment load in streams are one of the most important problems of the Himalaya (Marshall and Randhir 2008; Moore and Wondzell 2005; Poff et al. 2006; Rawat 1987). Preliminary studies conducted on this subject indicate that human interference, unscientific developmental activities, agriculture extension and unplanned way of road construction are some of the activities which are creating the hydrological imbalances in the region (Ives 1989; Singh and Pande 1989). In this region stream water yield both during rainy and non-rainy seasons has been found to be affected by both rainfall and geo-diversity of the spring recharge zone (Rawat and Rawat 1994; Negi and Joshi 2002). Impact of the recharge zone characteristics and land use, land cover changes on stream hydrology has also been reported in this region (Tiwari 2000; Rawat et al. 2011c; Rawat 2012). The Lesser Himalaya region is a densely populated non glacial tract of the Himalaya Mountain. Consequently perennial streams are fed by natural springs which are key source of water throughout the year.

The water sources of such springs and streams, in most cases are unconfined aquifers where the water flows under gravity. Spring water discharge fluctuations owe primarily due to rainfall pattern in the recharge area or more precisely stated, to variation in the amount of rainwater that is able to infiltrate the ground and recharge the ground water. The recharge of groundwater augments the discharge of the springs and seepages, and thus, causes more rapid outflow of the ground water. The high rate of discharge lowers the water table, reduces its gradient, and diminishes the pressure in pore spaces. This transience in recharge and discharge is the cause of the seasonal, local, and short-term fluctuations in spring and their stream water yield. The present study is an attempt to assess the geohydrological process of the springs using GIS technology. The Dabka catchment constitutes a part of the Kosi Basin in the Lesser Himalaya, India in district Nainital has been selected for the case illustration. This reconnaissance study estimates the spring discharge rate over the period of 6 years (2006–2011).The present work is an outcome of a multidisciplinary project on hydrological characteristics in Lesser Himalaya, sponsored by Department of Science and Technology (DST) Govt. of

India. In view of this, the present work is expected to be useful for both researchers and extension workers in the field of hydrology in general, and planning augmentation of water for drinking and other household consumption in particular.

2 Study Area

The study watershed lies between the latitude 29°24′09″–29°30′19″N and longitude 79°17′53″–79°25′38″E in the north-west of Nainital town near the tectonically active Main Boundary Thrust (MBT). The region encompasses a geographical area of 69.06 km^2 between 700 and 2,623 m altitude above mean sea level (Fig. 1). In most of the mountainous part of the watershed the surface gradient is steep and the topography is rugged.

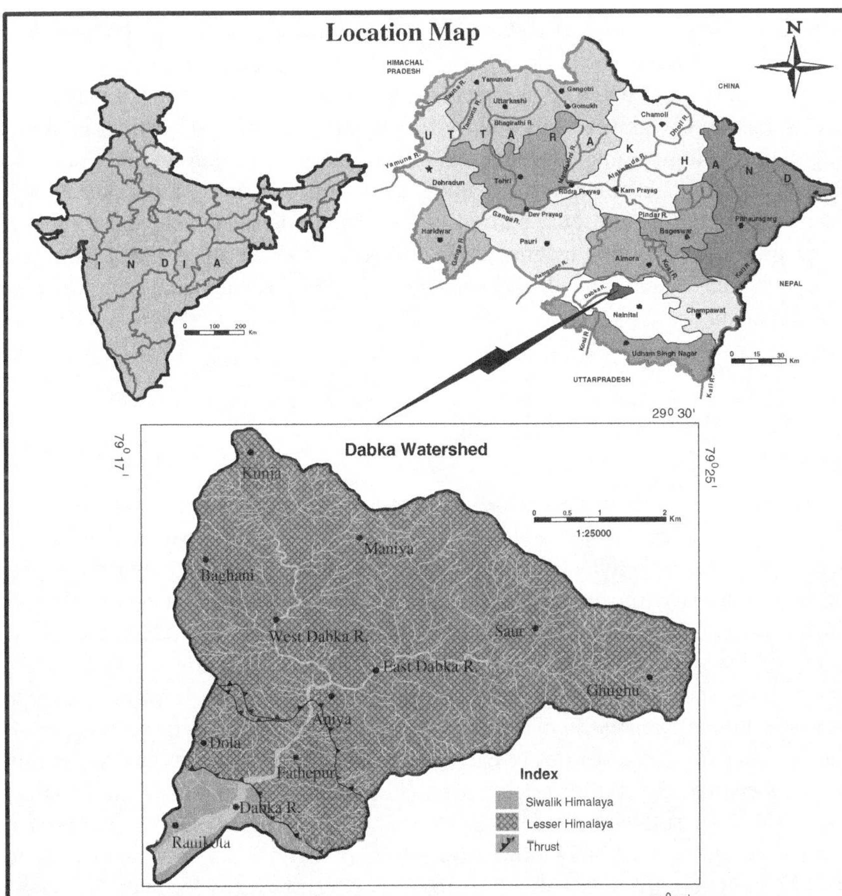

Fig. 1 Location map of study area

3 Methodology

The study comprises mainly two components, (a) lab study and (b) field investigations. Geo-structural maps were prepared during field study and details were verified and modified with other maps prepared during the lab study. The procedure adopted for the study has been discussed as below.

3.1 GIS Thematic Mapping

Survey of India toposheet published in 1986 on 1:25,000 scale (No53O/7, 53O/6) and Indian Remote Sensing Satellite (IRS-1C) LISS III and PAN merged data of two periods (1990 and 2010) were used for preparing base map employing visual and digital interpretation techniques. Digital interpretation was used for land use mapping. In order to overcome these constraints and also to attain the best possible level of accuracy in the interpretation, intensive ground truth surveys were carried out in the area and a visual interpretation key was evolved for primary land use classification. This was followed by the digital classification of final land use and geomorphological maps through on screen visual recording and rectification. Whereas, geology, geomorphology, lineament, slop gradient, slope aspect, soil type and drainage pattern etc., carried out by comprehensive field mapping following grid and isopleth's technique (Wentworth 1950; Strahler 1956) based on Survey of India Topographical Maps at scale 1:25,000 beside visual interpretation of satellite data.

3.2 Geo-hydrological Modeling

Geo-hydrological modeling comprises of comprehensive investigation about spring hydrology (i.e. types of springs, their geo-structural control, spatial distribution, density, discharge rates and changing pattern in relation to rainfall and land use pattern) and its impacts on stream discharge (i.e. monthly, seasonal, annual). All GIS thematic maps (i.e. geology, geomorphology, lineament, slope gradient and slope aspect, soil type, land use, drainage pattern etc.) superimposed to carryout spring types, density and their spatial distribution in the area whereas the water discharge monitored through representative sample spring approach. It is quite difficult to monitor water discharge data from each and every spring of the Himalayan terrain due the steep and rugged topography. It requires strategic planning and trained man power as well as sufficient funds; therefore representative sample spring approach of varied geo-ecosystem was followed for a 6 years (2006–2011) period. Out of the total 173 springs in the area only 42 springs have been selected for intensive monitoring of water discharge on monthly basis. These springs are representative of various types of springs in the basin. 7 sample springs were selected for each type of spring. The techniques used for discharge measurement according to the nature of the springs (i.e. running spring or dhara and pond or seepage). Dhara discharge was measured using a measuring cylinder and stop watch and seepage discharge was measured using gravimetric method (Rawat 1987):

$$S_q = \frac{S_a * W_d}{T} \times 1{,}000$$

S_q Rate of water discharge of Naula in l/m
S_a Surface area of Naula water in m^2
W_d Depth of water in m filled by seepage water during measurement
T Time taken in sec to fill up the original water surface.

4 Results and Discussion

4.1 Geo-environmental Setup

The geo-environmental setup of the study area consists of spatial distribution maps with their attribute data of different geo-sectors (i.e. geology, lineament, geomorphology, slope, soils, land use pattern and spatial distribution of landslides) and weather sectors (i.e. rainfall, temperature, humidity and evapo-transpiration loss). A brief description is given below.

4.1.1 Geology

Geologically the area is located in the southeastern extremity of the Krol belt forming outer part of Lesser Himalaya in Kumaun (Valdiya and Bartarya 1991). The watershed encloses rocks of the Blaini-Krol-Tal succession which are thrust over the autochthonous Siwalik Group along the Main Boundary Thrust (MBT) of Himalaya. The rocks of the area are divisible into Blaini and Krol groups (Pant and Goswami 2003; Rawat et al. 2011a). The Blaini Group has been further sub divided into Bhumiadhar, Lariakantha, Pangot and Kailakhan formations in an ascending order of succession (Fig. 2). The oldest rocks exposed in the watershed comprise quartzwacke, quartzarenite, diamictite, siltstone and shale (Bhumiadhar Formation) followed upward by predominantly arenaceous Lariakantha Formation, which inturn is followed by the diamictites, purple grey slates, siltstone and lenticular pink siliceous dolomitic limestone of the Pangot Formation.

The upper most Kailakhan Formation comprises dark grey carbonaceous pyritous slate and siltstone. The Blaini Group transitionally grades into the Krol Group. The lower most formation of the Krol Group is characterized by argillaceous marly sequence of the Lower Krol Formation (=Krol A). The Formation grades upward into purple green slates and yellow weathered dolomites with

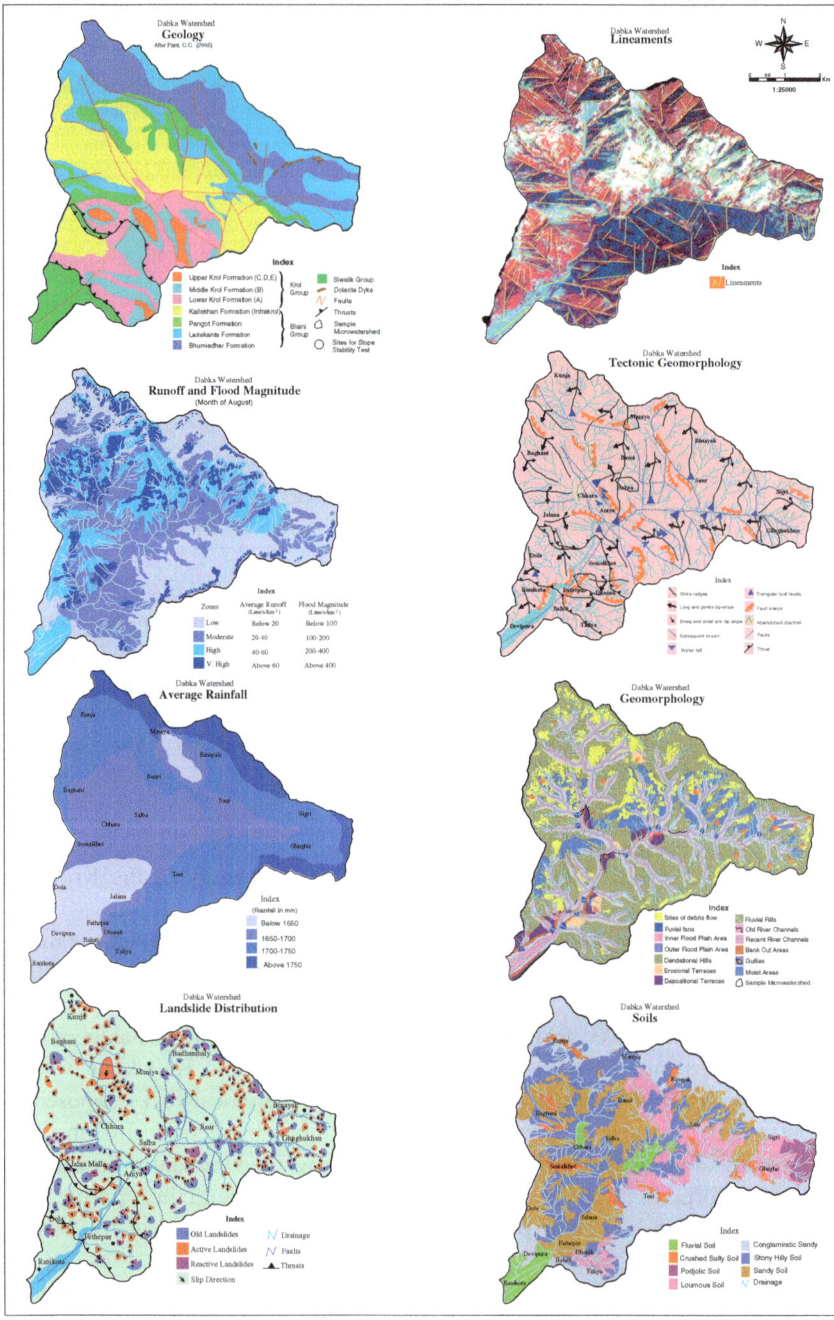

Fig. 2 Geo-environmental setup of study area

pockets of gypsum of the Hanumangarhi Formation (=Krol B). The Formation constitutes a marker horizon in the Krol belt. The Upper Krol Formation (Krol C, D, and E) is characterized by an assemblage of dolomitic limestone at the base followed by carbonaceous shales, fenestral dolomite showing cross bedding, brecciation and oolites and cryptalgal laminites. The upper most part is made up of massive stromatolitic dolomites locally cherty and phosphatic at places. The youngest Tal Formation comprises dolomite and carbonaceous slate, purple green slates interbedded with cross-bedded fine-grained sandstone and siltstone. The lower most southern part of the watershed comprises Siwalik Group made up of medium grained sandstone and shales.

4.1.2 Lineament and Structural Setting

A lineament is a linear feature in a *landscape* which is an expression of an underlying *geological structure* and controls the hazard vulnerability such as a *fault*, thrust etc. (Rawat et al. 2011c). Typically a lineament will comprise a fault-aligned valley, a series of fault or fold-aligned hills, a straight coastline or indeed a combination of these features. *Fracture zones*, *shear zones* and *igneous intrusions* such as *dykes* can also give rise to lineaments. Lineament orientations are dominantly found in NE to SW and NW to SE orientations in the area (Fig. 2).

4.1.3 Geomorphology

Geomorphologically, the watershed is constituted of as many as 27 types of landforms of different genetics, viz. fluvial, pluvial and tectonic. The fluvial landforms (i.e. landforms caused by the erosional and depositional processes of streams and rivers) in the watershed are low and middle level alluvial terraces, high level bedrock erosional terraces, small alluvial cones and fans, bank cut, narrow floodplains, sites of debris flow, moist areas, rills and gullies on alluvial terraces (Fig. 2). The pluvial landforms (i.e. landforms caused by erosional and depositional processes of rainwater) in the watershed are unchanneled concave valleys or hollows, knolls, cols, convex hill spurs (all erosional), and sheetwash cones and fans (depositional). Apart from these fluvial and pluvial landforms, the watershed is characterized by tectonic landforms (i.e. landforms caused by tectonic uplift) such as vertical rock and terrace scarps, abandoned valleys, waterfalls, rapids, soil and rock creeping zones, slumping zones, debris fall, rock fall and block gliding.

4.1.4 Slope

Four categories of slopes have been identified in the study area viz., gentle, moderate, steep and very steep slope (Fig. 2). The minimum parts of watershed having average surface slope less than 10° have been classified as gentle sloping areas.

About 4.79 % area of the watershed mainly along the master stream and its mouth has slope less than 10°. These areas constitute depositional landforms, such as, river terraces, debris fans etc. About 8.33 % area of the watershed has moderate slope from 10° to 20°. The convex and concave mid-crest slopes in Fatehpur and down slopes of Baluti village fall in the category of moderate surface slope. A considerably large proportion of the geographical area of the watershed (35.48 %) has steep surface slope varying from 20° to 30°. The steep slope areas constitute mainly landslide debris fans and their upslope areas in the western mountainous part of watershed, in villages Baghani, Jalna, Dola, Semalkhet, Saur, Gwalakhutti, Sigri and Hariyal. About 35.5 km^2 area that accounts for as much as 51.40 % of the total watershed area falls in very steep slope zone of greater than 30°. These very steep slopes are mainly occupied by different types of forests and barren and scrubland. All high altitudinal areas, mainly in the eastern part of the watershed have steep slope.

4.1.5 Soils

The main soils types found in the region include (i) Pebbly sandy soils which are mainly found in dense forests at higher elevations, (ii) Loamy soils have very coarse texture and composed of pebbles and cobbles which found mainly around Saur, Gwalakhuti, Dau, and in Bansi and Ghghukhan area of the watershed, (iii) Podzolic soils, which consist of high, humus contents extend over an area of 10 % of the catchment in Sigri, Binayak and Ghughukhan villages of the region, (iv) Crushed silty soil are seen in patches along the thrusts, faults and other structural feature, (v) Stony hilly soils are found around the upper slope of Bansi, Baghani and Dola-Jalna area, (vi) Fine sandy soils are found near Aniya and Chhara villages and in Fatehpur and Dhanak villages and (vii) Fluvial soil are found around Ranikota and Devipura areas in the watershed (Fig. 2).

4.1.6 Spatial Distribution of Landslides

A total of 248 landslides have been investigated in the area by comprehensive field work. Out of these 97 landslides are identified as active landslides, 91 identified as reactive and 83 being old landslides which area tectonically stable (Fig. 2). The stable or old landslides are further classified as recently stable/completely stable as per the field observation (Table 3). The landslide distribution map suggests that the active landslides were found in areas occupied by Lariakanta Formation of steep slopes. The major rock units of Krol Group, which are exposed in the vicinity of Main Boundary Thrust (MBT). Along the MBT the rocks are sheared, shattered, and therefore it is considered as a potential zone for slope failure especially during rainy season (Valdiya and Bartarya 1989, 1991). It was observed that dense forest have less density of landslides in the region. Tit Ka Danda area has lesser number of active landslides compared to other regions. This may be due to the presence of dense vegetation cover in the area (Fig. 2). But deforestation due to climate change and anthropogenic interference such as road construction extension of agriculture land, mining etc. has caused active slides in many places in the region. However, high runoff and denudation rate due to heavy rainfall has been a major factor in causing landslides in the weaker zones of the area in monsoon period. The reactivated landslides associated along the old stable landslides because of reactivation of old stable landslides due to accelerated land use dynamics, high runoff and erosion occurring by combination of anthropogenic and climate change factors (Rawat et al. 2011b).

4.1.7 Land Use Pattern

As mentioned in methodological section that the land use study consists of comparative land use land cover mapping for year 1991 and 2011 to assess the changes and identified the accelerating factors for the this changes as discussed below:

Land Use for Year 1991

On the basis of the above exercise, 7 land cover/land use classes could be identified in area, for 1991 and 2011. Besides the interpretation of the principal land use categories, all the major forest types available in the region could also be identified and mapped during the digital interpretation process. The land use/ land cover maps of the area have been prepared on the scale 1:25,000 and classified into 7 categories for both the period (1991 and 2011). The land use classes identified in the watershed include (i) Oak forests, (ii) Pine forests, (iii) Mixed forests, (iv) Scrub land, (v) Barren land, (vi) River beds and Water bodies, and (vii) Cultivated land which also includes settlements (Fig. 3). In 1991, out of the

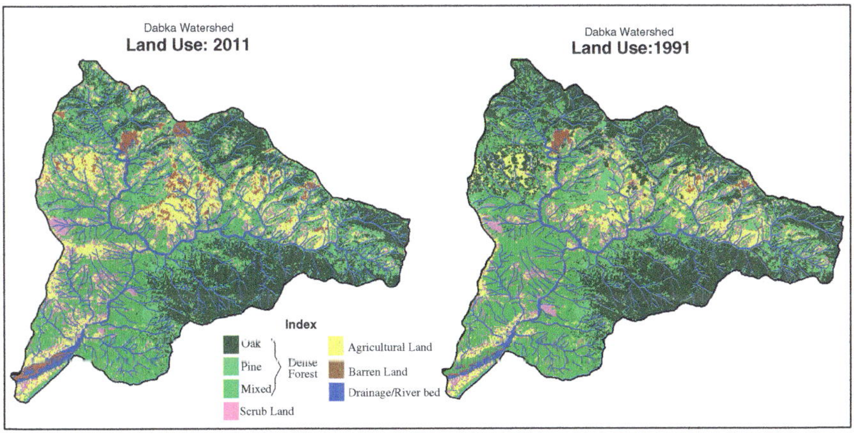

Fig. 3 Land use pattern of study area

total geographical area (69.06 km^2) of Dabka Watershed as much as 39.23 km^2 or 56.81 % was under forests.

The forests mainly include Oak forests, Pine forests and Mixed forests, which respectively account for 26.05 % (17.99 km^2), 11.89 % (8.21 km^2), and 18.87 % (13.03 km^2), of total geographical area of the watershed. The scrub land and barren land were respectively extended over 6.38 km^2 (9.24 %) and 2.18 km^2 (3.16 %) of the watershed area. Riverbeds and water bodies occupied only 1.50 km^2 or 2.17 % geographical area of the study region. Nearly 20 km^2 land surface which amounts to nearly 28.63 % of the total area of Dabka Watershed was under cultivation in 1991 (Table 1 and Fig. 3).

Land Use for Year 2011

The forest emerged as the major land use/land cover also in the year 2010. A geographical area of 36.77 km^2, which accounts for nearly 53 % of total area of the watershed, has been classified as forests. Due to complexities of terrain and other geomorphic features the forests of the watershed are diversified in nature. Out of the total forest 22.20 % (15.33 km^2) is under mixed forest, 19.56 % (13.51 km^2) is under Oak forest, and 11.48 % (7.93 km^2) is under Pine forests (Table 1). The hilly and mountainous parts of the watershed are covered with Oak and Pine species, whereas, in the lower elevations in the south, mixed type of vegetation is very common. Agriculture and settlement are now confined to 20.40 km^2 or 29.54 % of the total area. Scrub land, barren land and riverbeds and water bodies respectively extend over 6.22 km^2 (9.01 %), 3.39 km^2 (4.91 %), 2.28 km^2 (3.30 %) of the total geographical land surface of the area (Table 1 and Fig. 3).

Land Use Dynamics During 1991–2011

In order to monitor the dynamics of land transformation process land use interpretation was carried out for the years 1990 and 2010 using IRS LISS—III and PAN merged data for the respective years. The exercise revealed that oak and pine forests have decreased respectively by 25 % (4.48 km^2) and 3 % (0.28 km^2) thus bringing a decline of 4.76 km^2 forest in the watershed during 1991–2011. But, due to climate change the mixed forest taking place of oak forest in certain pockets and consequently the mixed forest in the catchment has increased by 18 % (2.3 km^2) during the same period which reduced the overall loss of forests in the region but its not eco-friendly as the oak forest because of the broad leaved and wide spread roots of oak trees helps in controlling the several hydrological hazards such as accelerated runoff, erosion, landslides, flash flood and river-line flood during monsoon period and drought during non-monsoon period. As a result, the watershed recorded a total decline of 2.46 km^2 or 6 % forest area during 1991–2011 (Table 1 and Fig. 3). The non-forest area has increased dramatically due to lopping and cutting of trees, accelerated runoff, soil erosion, and increased

Table 1 Land use pattern

Land use/land cover classes		1991		2011		Change (10 years)		Change (annual)		Accelerating factors for land use dynamics
		km^2	%	km^2	%	km^2	%	km^2	%	
Forest area	Oak	17.99	26.05	13.51	19.56	−4.48	−6.49	−0.45	−0.65	Climatic and anthropogenic
	Pine	8.21	11.89	7.93	11.48	−0.28	−0.41	−0.03	−0.04	Climatic and anthropogenic
	Mixed forest	13.03	18.87	15.33	22.20	2.3	3.33	0.23	0.33	Climatic only
	Total forest	39.23	56.81	36.77	53.24	−2.46	−3.56	−0.25	−0.36	Climatic and anthropogenic
Non-forest area	Scrub land	6.38	9.24	6.22	9.01	−0.16	−0.23	−0.02	−0.02	Anthropogenic only
	Barren land	2.18	3.16	3.39	4.91	1.21	1.75	0.12	0.18	Climatic and anthropogenic
	Riverbed	1.50	2.17	2.28	3.30	0.78	1.13	0.08	0.11	Climatic and anthropogenic
	Cultivated land	19.77	28.63	20.40	29.54	0.63	0.91	0.06	0.09	Anthropogenic only
	Total non forest	29.83	43.19	32.29	46.76	2.46	3.56	0.25	0.36	Climatic and anthropogenic
Total area		69.06	100.00	69.06	100.00	0.00	0.00	0.00	0.00	

agricultural activities. The non-forest area has mainly been confined to barren land, riverbed and cultivated land. Barren land increased 1.21 km^2 (56 %), riverbed increased 0.78 km^2 (52 %) and cultivated land increased about 0.63 km^2 (3 %) during the period of 1991–2011 (Table 1 and Fig. 3).

The results of land use dynamics presented on Table 1, advocates that the overall accelerating factor of land use dynamics in the study area broadly categorised as dominant factor and supporting factor. Out of the total seven classes of the land use land cover, five classes (i.e. Oak, Pine, Mixed, Barren and Riverbed) are being change dominantly due to climate change and anthropogenic factors plays a supporting role whereas only two classes (scrub land and agricultural land) are being change dominantly by anthropogenic and climate change factors plays a supporting role. Expansion of mixed forest land brought out due to upslope shifting of existing forest species due to climate change factor only because upslope areas getting warmer (Table 1 and Fig. 3).

4.1.8 Temperature

The average temperature of the Dabka Watershed is 19 °C whereas annual maximum and minimum temperature stand at 43 and 0 °C respectively (Table 2). The extreme maximum and minimum temperature records stand at 43 °C, in June on the south-facing barren land and 0 °C in January.

4.1.9 Rainfall

The watershed receives about 2,004 mm annual rainfall (Table 2). The average annual rainfall within the watershed varies between 1,623 mm and Maniya located in environmentally stressed barren hill slope and 2,187 mm at Ghughu located in dense forest hill slope. On other stations, the annual rainfall amount stand at 1,969 mm at Bansi located in agricultural land and 2,086 mm at Jalna located in fairly dense forest/shrub land (Table 2).

4.1.10 Humidity

The average annual humidity of the watershed stands at 83.4 % which varies between 86 % in the dense oak forest area and 64.5 % in the south-facing barren land. On other ecological conditions it stands at 80.7 % in fairly dense forest/shrub land, 78.5 % on the agricultural land area (Table 2).

4.1.11 Evaporation Loss

The actual annual evapotranspiration loss of the watershed is about 802.3 mm which approaches maximum up to 1,072.3 mm on the south-facing barren land

Table 2 Monthly average temperature, rainfall, humidity and evaporation in Dabka catchment (2005–2011)

Hydro-meteorological parameters	Monthly results												Annual Results
	January	February	March	April	May	June	July	August	September	October	November	December	
Temperature (in °C)	10.00	12.00	16.00	19.00	31.00	27.00	24.00	23.00	21.00	20.00	14.00	12.00	27.00
Rainfall (mm)	11.50	68.65	76.27	12.33	76.27	261.78	486.24	566.34	261.47	30.88	12.71	9.34	2000.00
Humidity (in %)	52.00	64.50	64.00	56.50	51.00	58.00	81.50	79.00	65.00	56.50	48.50	48.50	52.00
Evaporation rate (mm)	400.0	620.33	745.33	812.00	950.30	700.00	650.67	550.00	500.00	500.00	400.0	400.5	602.43

and drops down to 659.4 mm on the north-facing oak forest areas of the watershed. On other environmental conditions it was found 786.7, 811.4 mm on the fairly dense forest/shrub land, agricultural land areas respectively (Table 2).

4.1.12 Runoff and Flood Magnitude

The average runoff rate is 16.82 l/s/km^2 and it varies considerably depending upon the ecological conditions (Rawat et al. 2012a, b). It varies between 6.27 l/s/km^2 in the dense forest to 31.8 l/km^2/year on the barren land. On the fairly dense forest and agricultural land, the average rate of runoff stands at 11.05 and 16.98 l/s/km^2 respectively. These data advocate that the dense forested land due to broad leaved spices of trees has very high and the deforested barren land has very low water retention capacity within their hydrological system.

4.2 Spring Hydrology

The spring hydrology comprises of comprehensive investigation about types of springs, their geo-structural control, spatial distribution, density, discharge patterns etc. A brief description is presented below.

4.2.1 Spring Types and Their Spatial Distribution

There is a sum of parameters, such as, geological, hydrological, hydraulic, pedological, climatic and biological response that controls the origin of springs, which provides the basis for the classification of springs. Results suggesting three categories of springs according to their nature in the study area: (i) perennial springs having water flow throughout the year, (ii) non-perennial springs having water flow during some months mainly during rainy season, and (iii) dry springs. The spatial distribution mapping of spring has been carried out which depict that there are total 173 springs in the area out of which maximum 78 % (135 springs) being perennial, 16 % (28 springs) are non-perennial and minimum 6 % (10 springs) are categorized as dried up springs (Fig. 4). The spatial distribution maps also suggest four categories of spring density in the area i.e. low, moderate, high and very high (Fig. 4).

4.2.2 Geo-structural Controls of Springs

Further all three categories of springs, can be classified into seven different types according to their controlling factors (Fig. 5), these are: (i) thrust controlled, (ii) fault controlled, (iii) bedding controlled, (iv) fracture controlled, (v) joint controlled, and (vi) shear zone controlled springs (vii) fluvial deposits related springs.

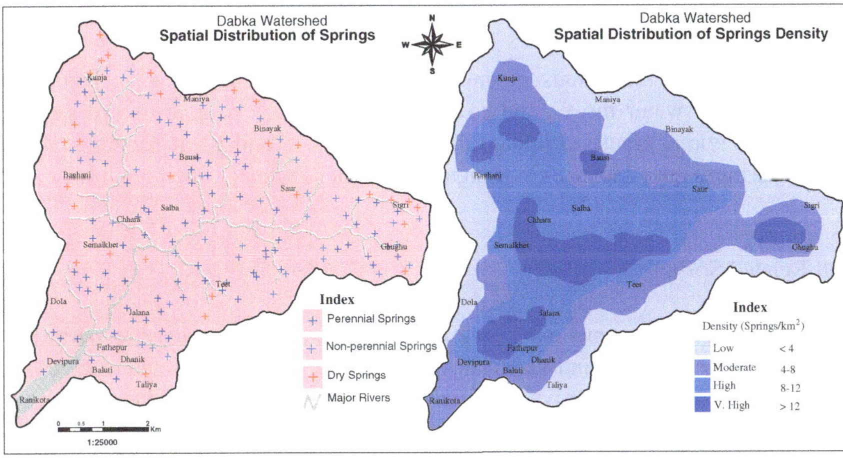

Fig. 4 Spatial distribution of springs

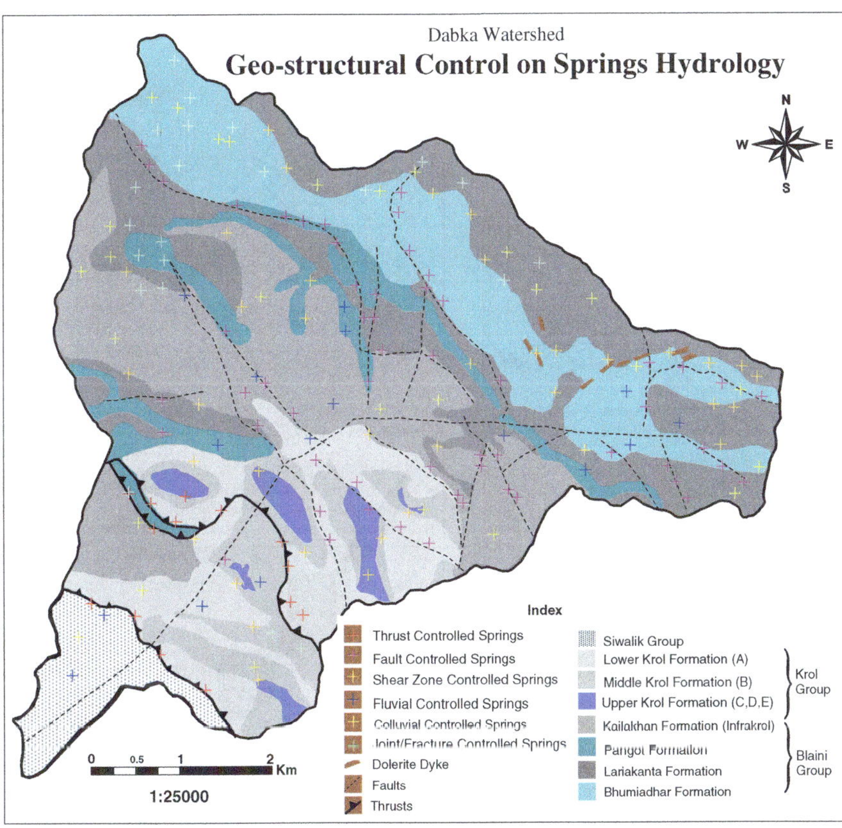

Fig. 5 Geo-structural controls of springs

Thrust Related Spring

Thrust planes and lineaments extending deep underground serve as pathways along which the water moves and rises up to the surface to give rise to springs such as in Bagjala, Gautiya, Jalna Talla areas (Figs. 5 and 6). It accounts about 8 % (14 springs) out of total 173 springs in the area (Table 3).

Fault Related Spring

Fault planes and lineaments extending deep underground also serve as pathways along which the underground water moves and rises up to the surface to give rise to springs such as seen in Aniya, Chhara, Dau, Salwa, Bansi areas (Figs. 5 and 6). It accounts maximum about 33 % (57 springs) out of total 173 springs in the area (Table 3).

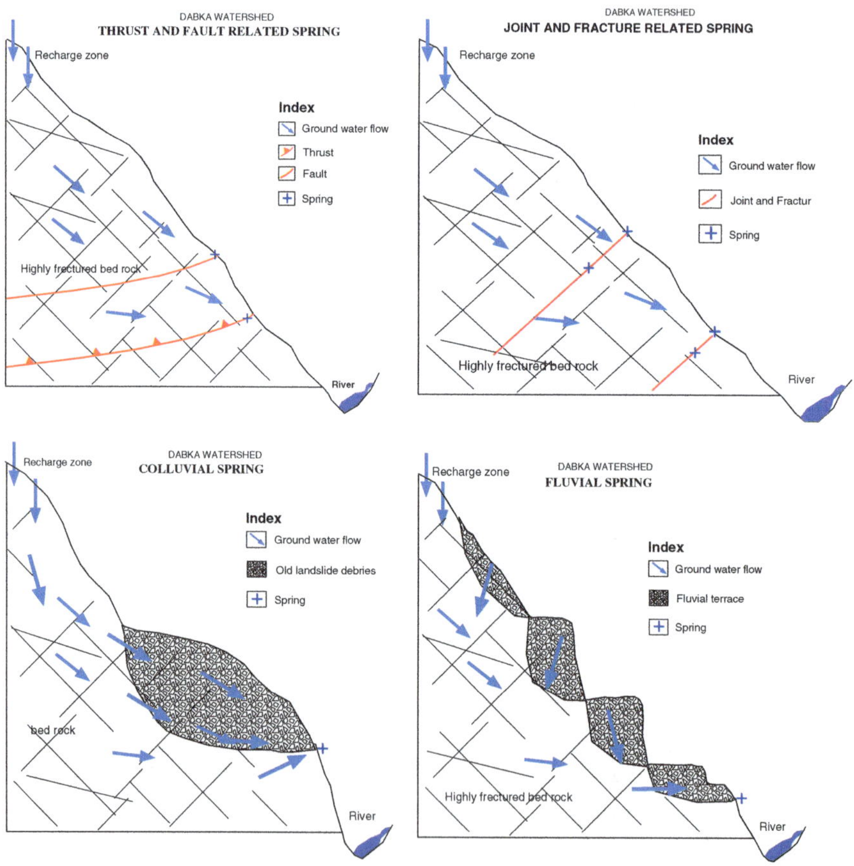

Fig. 6 Schematic diagrams of different types springs in Dabka catchment

Table 3 Geo-structural controls of springs

Exiting types springs	Total number of springs	% of total springs
Thrust related springs	14	8
Fault related springs	57	33
Shear zone controlled	37	21
Fluvial springs	16	9
Coluvial springs	27	16
Joint/fracture related springs	22	13
Total	173	100

Joint and Fracture Related Springs

Springs of this type are formed in fractured hard sedimentary rocks, such as, in Jawa, Ranikota, Jalna areas of the basin. These springs originate either along the hills slopes or along streambeds wherever the local water table is intersected by fractures or by the ground surfaces (Figs. 5 and 6). It accounts about 13 % (22 springs) out of total 173 springs (Table 3).

Colluvial Springs

The springs formed with in colluvial deposits and receive water from upslope along the fault, fracture and joints are known as colluvial springs. The upper slopes of recharge areas are generally low (8°–12°) which provide good hydraulic gradient. Locally, there is an overlapping of conditions of colluvial and fracture related springs where the spring discharge of the colluvial deposits also receive groundwater flowing down along the fracture and joints (Figs. 5 and 6). The examples of such conditions may be seen in Baghni, Maniya, Chhara, Ghughukhan, Saur, Fathehpur, and Jalna areas. It accounts about 16 % (27 springs) of total 173 springs (Table 3).

Fluvial Deposits Related Springs

Because of the difference in porosity and permeability of the deposits and underlying bedrock, the groundwater is discharged as springs through the contact of alluvial fans and terrace deposits on terraces and also at the contact of fluvial deposits and the bedrocks (Figs. 5 and 6). The fluvial deposit related springs are mainly found in Ranikota and Devipura areas of the region. It accounts about 09 % (16 springs) out of total 173 springs (Table 3).

Shear Zone Related Springs

Due to overlay of different lithology along thrusts, shear zones and faults at several places in the area such types of springs are produced. This type of springs accounts about 21 % (37 springs) out of total 173 springs. Such springs are seen near village Bansi, Salba, Aniya, Dau. (Figs. 5 and 6).

4.2.3 Spring Discharge

A comprehensive study of all 42 representative sample springs and their water discharge pattern has been carried out through hydrograph analysis, which shows the state of discharge, velocity and other properties of water flow with respect to time (Fig. 7). Hydrograph consists of three parts, these are approach segment, rising segment and recession segment. In the approach segment the water flow approaches minimum, in the rising segment the water discharge rises, and the period of decrease in water discharge is known as recession or falling segment of the hydrograph. The point of maximum water discharge is known as hydrograph peak. The upper points of the rising and recession segment are known as hydrograph crest. Although all 42 sample springs have their own discharge capacity with their controlling factors but above three segments of spring discharge (approach, rising and recession segments)

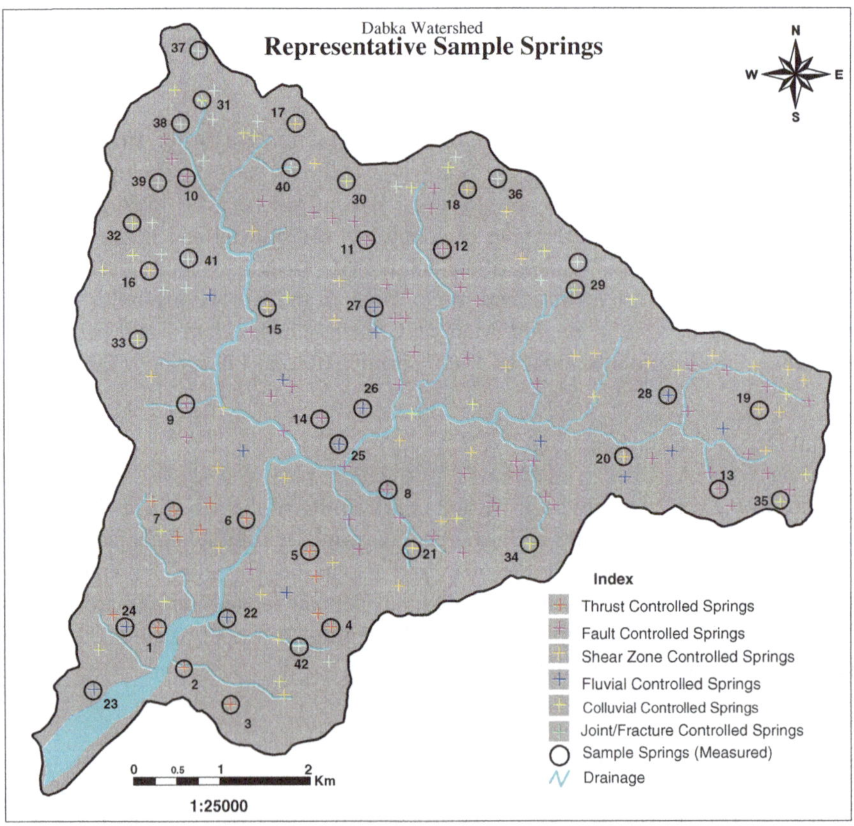

Fig. 7 Representative sample springs selected for comprehensive study on spring hydrology

varies in the entire watershed due to varying meteorological conditions, particularly of the pattern of precipitation and geological conditions as discussed below.

Annual Average Discharge

It is 6 years (2006–2011) average discharge. The annual average discharge stands about 22 l/min in the area but it varies in respect to geo-environmental controlling factors of the springs (Fig. 8 and Table 4). The hydrograph analysis suggests that:

- Thrust and fault controlled springs having dense vegetation cover of oak forest in the upslope which constitutes highest average discharge with the rate of respectively 39 and 32 l/min (Fig. 8 and Table 4).
- Joint and fracture controlled springs having rocky barren land in the upslope constitutes lowest average discharges rate i.e. 10 l/min (Fig. 8 and Table 4).
- Coluvial, fluvial and shear zone controlled springs constitutes moderate to high average discharge, accordingly extent of vegetation cover in the upslope (Fig. 8 and Table 4).

Monthly Average Discharge

It has been observed that the monthly averages discharge varies according to the precipitation rate during the period 2005–2011 (Table 5). The hydrograph analysis suggesting that:

- The period of the rising segment varies from year to year mainly depending upon rainfall pattern (Table 5).

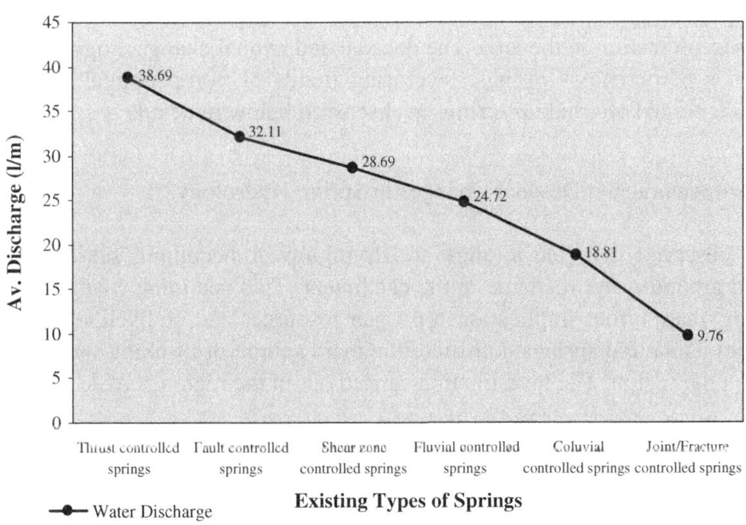

Fig. 8 Annual average discharge of springs

Table 4 Annual average discharge

Exiting types springs	Total number of springs	% of total springs	Number of Sample springs	Average annual discharge in l/min
Thrust related springs	14	8	07	38.69
Fault related springs	57	33	07	32.11
Shear zone controlled	37	21	07	28.69
Fluvial springs	16	9	07	24.72
Coluvial springs	27	16	07	18.81
Joint/fracture related springs	22	13	07	9.76
Total	173	100	42	152.78

- On an average the water discharge starts rising since the month of May (Table 5).
- April and December constitute the approach (base flow) segment of the hydrograph (Table 5).
- August constituted the rising (peak flow) segment of the hydrograph (Table 5).
- January, February and March constitute the recession segment of the hydrograph (Table 5).

4.2.4 Dynamics in Spring Hydrology and Its Impact on Underground Water and Stream Flow

The present results compared with previous studies have been carried out in the area and also in adjoining catchment. Year 1991 determined as a base year for previous status of spring hydrology as the same year have been determined for land use change detection in the area. The decadal and annual changes suggest decreasing trends of perennial springs, increasing trends of non-perennial springs and increasing trends of dried up springs as discussed below in detail.

Trends of Annual and Decadal Changes in Spring Hydrology

It was observed that, the springs are drying-up or becoming seasonal due to reduced ground water recharge in the catchment. This declining change in stream hydrology has serious implications on water resources and on livelihood and food securities as natural springs constitute the main source of drinking water and irrigation in the region. The investigation carried out in the region revealed that:

- Annually 3 perennial springs are being converted into non perennial or dried up springs as decadal and couple of decadal changes suggests respectively 25 and 51 perennial spring's reduction (Fig. 9 and Table 6).
- Non-perennial springs are increasing with a rate of 1 spring/year and 14 springs/decade (Fig. 9 and Table 6).

Table 5 Monthly average spring discharge under different types of geohydrology in Dabka watershed

Exiting types springs	January	February	March	April	May	June	July	August	September	October	November	December
Thrust related springs	8.73	9.28	10.37	4.74	8.65	68.90	93.66	133.80	80.28	27.30	10.92	7.64
Fault related springs	7.21	7.66	8.56	1.60	9.29	57.20	77.28	111.40	67.24	22.52	9.01	6.31
Share zone controlled	6.46	6.87	7.67	2.30	7.83	51.50	69.30	99.00	59.40	20.20	8.08	5.65
Fluvial springs	5.56	5.91	6.60	2.10	6.76	44.60	59.64	85.20	51.12	17.38	6.95	4.87
Coluvial springs	4.23	4.49	5.02	1.50	4.81	34.40	45.36	64.80	38.88	13.22	5.29	3.70
Joint/fracture related springs	2.19	2.33	2.60	0.57	1.89	18.80	23.52	33.60	20.16	6.85	2.74	1.92
Rainfall (mm)	11.50	68.65	76.27	12.33	76.27	261.78	486.24	566.34	261.47	30.88	12.71	9.34

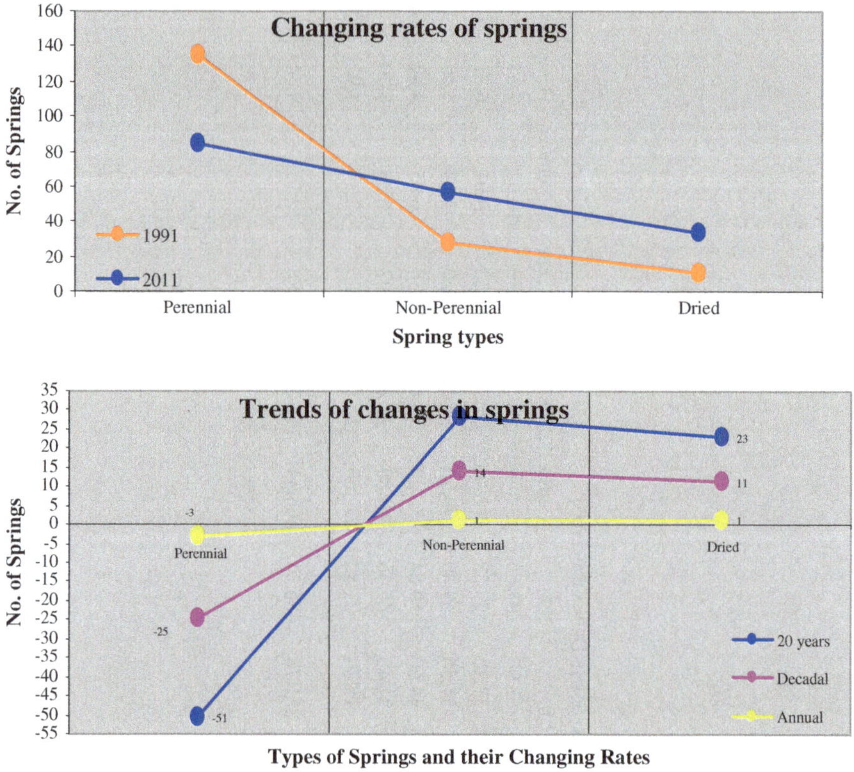

Fig. 9 Trends and annual and decadal changes in spring hydrology

Table 6 Trends of annual and decadal changes in spring hydrology

Spring types	1991		2011		Changes		
	No.	%	No.	%	years	Decadal	Annual
Perennial	135	78	84	49	−51	−25	−3
Non-perennial	28	16	56	32	28	14	1
Dried	10	06	33	19	23	11	1
Total	173	100	173	100			

- Dried springs are also increasing with a rate of 1 spring/year and 11 springs/decade (Fig. 7 and Table 6).
- Accelerated land use change is a major cause for the degradation of spring hydrology (Fig. 9 and Table 6).

Declining Underground Water

Land use degradation and deforestation has reduced the protective vegetal cover as a result significant proportion of rainfall goes waste as flood water without replenishing the groundwater reserves. It is noticed that the ground water level in general is gradually going down mainly due to deforestation and high flood runoff. It has been observed that the water which was easily accessible at 2,000 m altitude has gone down up to just 1,200 m altitude (Fig. 10). The decreasing trends of underground water label has affected the spring hydrology and stream hydrology in the area as also depicted by a schematic diagram in Fig. 9 for year (1991–2011).

Decreasing Trends of Stream Flow

As a consequence of drying up of natural springs the streams are also depicting decreasing trends of annual stream discharge and becoming perennial to seasonal streams in the area. The existing average annual discharge of Dabka Master River is 11 l/s/km^2 whereas during 1985–1991 it was quit high i.e. 17 l/s/km^2. So the stream flow has been decreased about 35 % during last two decades. It is also found that due to decreasing trends of stream discharge a number of perennial streams have gone dry.

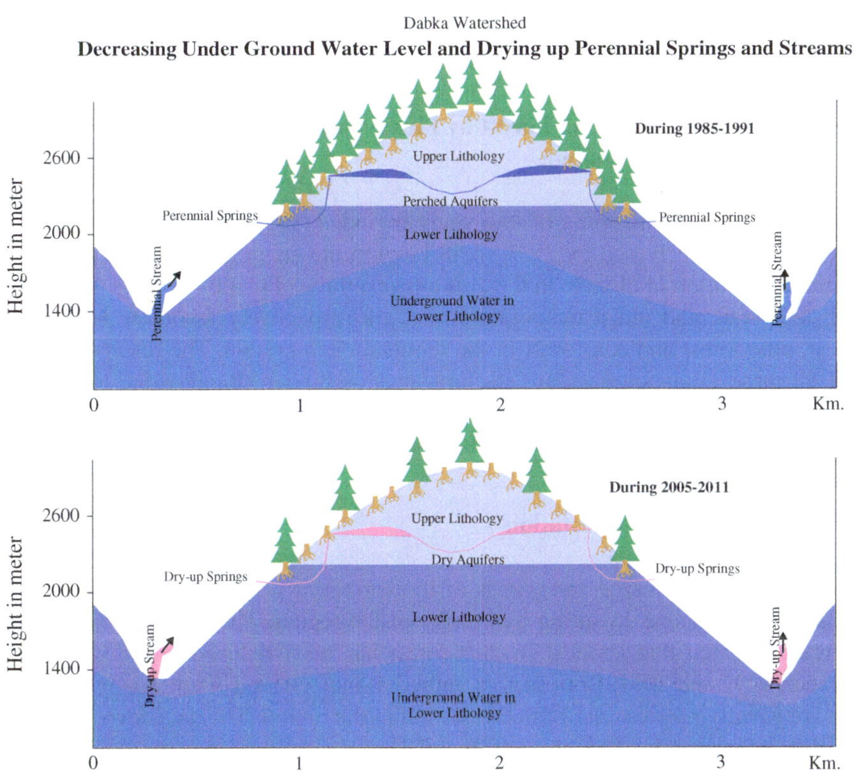

Fig. 10 Declining underground water

4.3 Major Responsible Factors for Declining Changes in Spring Hydrology

It has been observed that a number of natural environmental and anthropogenic factors affected the geo-hydrology of springs in Dabka catchment. The major natural environmental factors identified are decreasing rainfall perhaps due to global climate change, replacement of oak forest by pine forest, deforestation, neo-tectonic movements, landsides etc. The major anthropogenic factors identified are encroachment of forest for agricultural purposes, road construction, upslope cutting for settlements and big buildings. A brief description of each responsible factor given in Fig. 10.

4.3.1 Climate Change and Decreasing Annual Average Precipitation

It is well established through several studies that annual rate of average rainfall has decreased due to climate impacts in Himalaya (Shrestha 2009; Singh et al. 2011). The decreasing precipitation rate leads to water deficiency problem in spring sanctuaries in the region (Rawat et al. 2012a, b). Consequently, it has been observed in the area that the number of perennial springs is being converted into non-perennial springs and also drying up due to water deficiency (Rawat et al. 2011b).

4.3.2 Encroachment of Oak Forest by Pine Forest Due to Climate Change

GIS based land use mapping of two study periods (1991 and 2011) revealed that oak forests has decreased by 25 % (4.48 km^2) in the area during 1991–2011. The spatial distribution of climate in the area suggest three climatic zones i.e. sub-tropical, temperate and moist temperate which are respectively favorable for mixed forest, pine forest and oak forest in the mountain eco-system. The results indicate that all these climatic zones are shifting towards higher altitudes due to climate change and affecting the favorable conditions of the existing land use pattern and reduction of the oak forest and its replacement pine and mixed forest.

4.3.3 High Rate of Deforestation

Deforestation in recharge zone areas of the springs is a major responsible factor for declining changes in spring hydrology of the region. The GIS based land use exercise revealed that about 4.76 km^2 forest has been reduced in the watershed between 1991 and 2011. Both factors, either climate change or anthropogenic factors are equally responsible for deforestation in the region. The data advocates that the dense forested land comprising broad leaved spices of trees have very high and the deforested barren land has very low water retention capacity within their hydrological system.

4.3.4 Expansion of Agricultural Land

Non-scientific agricultural practices in highly vulnerable slopes and encroachment into forest land is also found to be a major factor for decreasing trends of spring water discharge. During last 2 decades (1991–2011) the agricultural land has expanded about 3 % (0.63 km^2) as carried out by GIS land use mapping using remote sensing data of 1991 and 2011.

4.3.5 Increased Hydrological Hazards

It has been demonstrated by several studies that the climate change, decreasing rainfall, high rate of land degradation, deforestation etc. are accelerating several hydrological hazards during monsoon (i.e. extreme rainfall events, high runoff, river line flood, flash flood, soil erosion and landslides etc.) period and non-monsoon period (water deficiency and drought hazard etc.). Consequently it is reducing percolation of rain water.

4.3.6 Neo-tectonic Movements

The area located in proximity of the Main Boundary Thrust (MBT) which is seismotectonically active. It has been observed by tectonic geomorphological investigations in the Dabka catchment there are neo-tectonically active thrust, faults and other lineaments. Consequently the flow of underground water is being affected.

4.3.7 Slope Failures and Landslides

There are a number of geological features which cause slope failure and landslide hazards in the area (Rawat et al. 2011b). As already discussed, that the area is seismically very active and has been experiencing seismic movement along the thrusts and faults causing slope failure. The slope failure and landside have influenced the spring hydrological process.

4.4 Need for Integrated Engineering and Land Use Measures for Sustainable Development of Spring Sanctuaries

The geo-hydrology of springs in Dabka basin has been affected by several natural environmental and anthropogenic factors. Therefore, there is an urgent need of protecting the springs by creating spring sanctuaries and makes it necessary to adopt an integrated and comprehensive successful engineering and land use measures based on the scientific interpretation of the crucial linkages among land use, climate change and spring hydrology.

As first step towards the management of spring sanctuaries, a comprehensive land use plan has been evolved and proposed for Dabka Watershed which could be implemented in the Himalayan regions because of its being eco-friendly and low cost effective. The land use framework has been designed taking into consideration of geo-parameters (geomorphology, slope, drainage density, drainage frequency, soils, geology and land use pattern), hydro-parameters (comprising of temperature, rainfall, humidity, evapotranspiration loss, runoff, sediment transport etc.) and hazard-parameters (drought, flood, erosion and landslides). The land use plan evolved for the region has been presented in Table 7 and described as under:

(i) The conservation, protection and sustainable development of forest resources are essential for the management of spring sanctuaries. In view of this, the forest area of the Dabka watershed has been proposed to be increased from the existing 36.56–45.00 km^2. This increase would be possible by bringing additional 8.22 km^2 of lopped-forest area or shrubs land and agricultural land (having high vulnerability for hydrological hazards) under reforestation programme (Table 7). The lopped-forest areas under barren, shrubs and agricultural land proposed to be afforested is already a part of forest, but it is under highly degraded condition (Table 7).
(ii) Out of total (20.51 km^2) existing agricultural land, 16 % (4.31 km^2) which are under high hydrological hazards has been proposed to be converted to shrubs land and/or horticultural practices (Table 7).
(iii) The favourable geographical conditions for the production of a variety of fruits in the watershed, and community demand for horticultural development, 5.48 km^2 area of existing barren, shrubs and cultivated land which have high risk of hydrological problem has been recommended to be brought under horticultural development in the watershed (Table 7).
(iv) As much as 2 km^2 area under existing barren and cultivated land has been suggested to be developed as shrub land (Table 7).

Table 7 Proposed land use as an eco-friendly measure for the development of spring sanctuaries

Land use classes	A. Existing area		B. Proposed area	
	km^2	%	km^2	%
1. Dense forest (Oak, Pine and mixed)	36.77	53	45.00	65
2. Shrubs land	06.22	09	02.00	03
3. Barren land	03.39	05	00.00	00
4. Cultivated land	20.40	30	16.30	24
5. River bed	02.28	03	2.28	03
6. Horticulture	00.00	00	3.48	05
Total	69.06	100	69.06	100

(v) Besides these, some engineering measures such as rainwater harvesting, construction of check dams, wall, fence, infiltration tanks gabions etc. be implemented for sustainable development of spring sanctuaries.

5 Conclusion

The study demonstrates that the springs being controlled by geological structure, land use pattern and extent of vegetation cover in their upslope areas. Consequentially the thrust and fault controlled springs having dense vegetation cover of oak forest in their upslope areas have highest average discharge. Joint and fracture controlled springs having rocky barren land in the upslope manifest lowest average discharges. The month of August witness the peak flow of the spring discharge which shows positive correlation with rainfall pattern. The month of April constitutes the base flow of the spring discharge because of driest month of the year. Reduction in the protective vegetal cover has resulted in significant proportion of rainfall as surface runoff without replenishing the groundwater reserves. Consequently, the perennial springs have decreased at a rate of 3 springs/year whereas the non-perennial and dried up springs increased at the rate of 1 spring/year. This has resulted in acute drinking and irrigation water paucity in area. Therefore there is an urgent need of conserving the existing springs and revving/rejuvenation of dried up spring and creating spring sanctuaries. It is hoped that the study would help the village community, district and state development authorities to formulate Decision Support System (DSS) for planning and better management of the degraded watersheds in the Himalaya region.

Acknowledgments This study constitutes part of a multidisciplinary project, "*Geo-environmental Appriasal of the Dabka Watershed, Kumaun Lesser Himalaya, District Nainital: A Model Study for Sustainable Development*" *by* Department of Science and Technology (DST) Govt. of India, (Project *No. ES/11/599/01 Dated 27/05/2005*). The authors are thankful to Dr. Nisha Mendiratta for the support and guidance during the course of investigation. Thanks to also due to Shri M.S. Bargali, Project Assistant for helping during the intensive field work.

References

Bartarya SK (1991) Watershed management strategies in Central Himalaya. The Ganga River Basin, Kumaun, India. Land Use Policy 8(3):177–184

Bisht BS, Tiwari PC (1996) Land use planning for sustainable resource development in Kumaon Lesser Himalaya—a study of Gomti watershed. Int J Sustain Dev World Ecol 3:23–34

Davies-Colley RJ, Smith DG (2001) Turbidity, suspended sediment and water clarity: a review. Am Water Resour Assoc 37(5):1–17

Dunn SM, Lilly A (2001) Investigating the relationship between a soils classification and the spatial parameters of a conceptual catchment-scale hydrological model. Hydrology 252(1–4):157–173

Haigh MJ, Rawat JS, Bisht HS (1989) Hydrological impact of deforestation in Central Himalaya. In: Proceedings hydrology of mountainous areas, Czechoslovakia, IHP/UNESCO, pp 425–426

Hundecha YH, Bárdossy A (2004) Modeling of the effect of land use changes on the runoff generation of a river basin through parameter regionalization of a watershed model. Hydrology 292:281–295

Ives JD (1989) Deforestation in the Himalaya: the cause of increased flooding in Bangladesh and Northern India. Land Use Policy 6:187–193

Marshall E, Randhir TO (2008) Effect of climate change on watershed processes: a regional analysis. Clim Change 89:263–280

Moore RD, Wondzell SM (2005) Physical hydrology and the effects of forest harvesting in the Pacific Northwest: a review. American Water Resources Association 41:763–784

Nearing MA, Jetten V, Baffaut C, Cerdan O, Couturier A, Hernandez M, Le Bissonnais Y, Nichols MH, Nunes JP, Renschler CS, Souchère V, Oost K (2005) Modeling response of soil erosion and runoff to changes in precipitation and cover. Catena 61(2–3):131–154

Negi GCS, Joshi V (2002) Drinking water issues and development of spring sanctuaries in a mountain watershed in the Indian Himalaya. Mt Res Dev 22(1):29–31

Pant CC, Goswami PK (2003) Tide-storm dominated shelf sequence of the neoproterozoic Blaini formation and its implications on the sedimentation history of Krol belt, Kumaun Lesser Himalaya, India. Nepal Geol Soc 28:19–39

Poff NL, Bledsoe BP, Cuhaciyan CO (2006) Hydrologic and geomorphic alteration due to differential land use across the United States. Geomorphology 79:264–285

Randhir TO (2005) Managing ecosystems in the presence of habitat interactions and market imperfections in a dynamic setting. Int J Ecol Econ Stat 3(5):21–41

Rawat JS (1987) Morphology and morphometry of the Naini Lake, Kumaun Lesser Himalaya. J Geol Soc India 30:493–498

Rawat JS, Rawat MS (1994) Accelerated erosion and denudation in the Nana Kosi Watershed, Central Himalaya. Part I: sediment load. Mt Res Dev 14(1):25–38

Rawat PK (2012) Impacts of climate change and hydrological hazards on monsoon crop pattern in Lesser Himalaya: a watershed based study. Int J Disaster Risks Sci 3(2):98–112

Rawat PK, Pant CC, Tiwari PC (2011a) Morphometric analysis of third order river basins using high resolution satellite imagery and GIS technology: special reference to natural hazard vulnerability assessment. Int Sci Res J 3(2):70–87

Rawat PK, Tiwari PC, Pant CC (2011b) Climate change accelerating hydrological hazards and risks in Himalaya: a case study through remote sensing and GIS modeling. Int J Geomatics Geosci 1(4):678–699

Rawat PK, Tiwari PC, Pant CC (2011c) Modeling of stream runoff and sediment output for erosion hazard assessment in Lesser Himalaya; need for sustainable land use plan using remote sensing and GIS: a case study. Int J Nat Hazards 59:1277–1297

Rawat PK, Tiwari PC, Pant CC (2012a) Geo-hydrological database modeling for integrated multiple hazard and risks assessment in Lesser Himalaya: a GIS based case study. Int J Nat Hazards 62(3):1233–1260

Rawat PK, Tiwari PC, Pant CC, Sharama AK, Pant PD (2012b) Spatial variability assessment of river-line floods and flash floods in Himalaya: a case study using GIS. Int J Disaster Prev Manag 21(2):135–159

Shrestha AB (2009) Climate change in the Hindu Kush-Himalayas and its impacts on water and hazards. Asia Pac Mt Netw Bull 9:1–5

Singh AK, Pande RK (1989) Changes in the spring activity: experiences of Kumaun Himalaya, India. The Environ 9(1):75–79

Singh P, Bengtsson L (2004) Hydrological sensitivity of a large Himalayan basin to climate change. Hydrol Process 18:2363–2385

Singh SP, Khadka I, Bassignana Karky, B.S. and Sharma, E. (2011) Climate change in the Hindu Kush-Himalayas: the state of current knowledge. International Centre for Integrated Mountain Development, Kathmandu

Strahler AN (1956) Quantitative slope analysis. Bull Geol Soc Am 67:571–596

Sutherland AB, Meyer JL, Gardiner EP (2002) Effects of land cover on sediment regime and fish assemblage structure in four southern Appalachian streams. Freshw Biol 47(9):1791–1805

Tiwari PC (2000) Land use changes in Himalaya and their impact on the plains ecosystem: need for sustainable land use. Land Use Policy 17:101–111

Valdiya KS, Bartarya SK (1991) Hydrogeological studies of springs in the catchment of Gaula River, Kumaun Lesser Himalaya, India. Mt Res Dev 11(3):239–258

Valdiya KS, Bartarya SK (1989) Problem of mass-movement in part of Kumaun Himalaya. Curr Sci 58:486–491

Wentworth CK (1950) A simplified method of determining the average slope of land surfaces. Am J Sci 5:20

Chemical Characterization of Meltwater from East Rathong Glacier Vis-à-Vis Western Himalayan Glaciers

Brij M. Sharma, Shresth Tayal, Parthasarathi Chakraborty and Girija K. Bharat

Abstract Ice and meltwater samples from Rathong glacier and its pro-glacial stream Rathong chu, were collected during late winter season. Samples were analyzed to study the chemical composition, weathering, and geochemical processes in ice and meltwater at high altitudes. This study elucidate the variation of major cations (Li^+, Mg^{2+}, NH_4^+, K^+, Na^+, Ca^{2+}) and major anions (PO_4^{3-}, F^-, NO_2^-, Cl^-, SO_4^{2-}, NO_3^-) with respect to ionic concentration reported for glaciers in western Himalaya High correlation between Ca^{2+} and Mg^{2+}, the contribution of ($Ca^{2+} + Mg^{2+}$) to the total cations and high value of $[(Ca^{2+} + Mg^{2+})/(Na^+ + K^+)]$ ratio suggests that the major source of dissolved ions in the meltwater is carbonate weathering. Enrichment of samples with NO_3^- and NH_4^+ suggests that scavenging of HNO_3 present in the atmosphere is a major contributor for these ions. High average value of Na^+/Cl^- and K^+/Cl^- ratio suggest that marine aerosols are not the only contributor to ionic composition. R-mode factor analysis was performed to understand the contribution of geochemical weathering and atmosphere in accumulating major cations and anions in ice and meltwater, which suggests that a majority of ions were from chemical weathering and post-depositional sources.

Keywords Chemical composition · Rathong glacier · Pro-glacial stream · Meltwater · Chemical characterization · Major ions · Western Himalaya

B.M. Sharma
TERI University, Vasant Kunj, New Delhi, India

S. Tayal (✉) · G.K. Bharat
TERI, India Habitat Centre, New Delhi, India
e-mail: stayal@teri.res.in

P. Chakraborty
Geological Oceanography Division, National Institute of Oceanography, Goa, India

R. Joshi et al. (eds.), *Dynamics of Climate Change and Water Resources of Northwestern Himalaya*, Society of Earth Scientists Series,
DOI 10.1007/978-3-319-13743-8_14

1 Introduction

Glaciers and Polar Regions have obtained the attention of scientific community in the last few decades, not only because they play an important role in global hydrological cycle but also for providing significant environmental information of climate change and biogeochemical processes in the surrounding atmosphere. Non-polar regions and mountain glaciers (an elevation greater than 4,000 m above sea level (m.a.s.l.) provide immense amount of information on the century or even higher time scale about the atmospheric constituents like mineral dust, anthropogenic pollution, weathering and geochemical processes controlling the water and sediment chemistry in high altitude rivers (Wagenbach 1989).

Glacier snow and ice have the property to accumulate pollutants, sea salts, acids and other chemical compounds and later release them into atmosphere through air-water gas exchange or during snowmelt runoff (Krnavek et al. 2012). Glaciers not only act as accumulation site for different chemicals but also are very influential agents for erosion and contribute substantial amount of sediments as well as solutes to river system (Hasnain et al. 1989; Collins and Hasnain 1995). The chemical characteristics of meltwater discharged from the glacier are different from other aqueous systems in terms of chemical reactions. Meltwater from the snout of a glacier is enriched in numerous chemicals (Drewry 1986). This is due to effective hydro-geochemical reactions taking place within the glacier. Chemical species carried by proglacial meltwater are derived from the supraglacial, englacial, and subglacial zones. Preferentially, prolonged contact between water and sediments result in chemical weathering processes, while the existence of supplies of readily mobilized ions maximizes solute uptake (Collins 1979; Tranter et al. 1993; Sharp et al. 1995; Brown et al. 1994, 1996). The Himalayan region represents an ideal environment for studying hydro geochemical processes of remote area because of its geographical location and influence on the global atmospheric circulation. The main river systems which originate from the Himalayan glaciers are Indus, Ganga and Brahmaputra (Puri 1994). 60–70 % of the fresh water to these river systems is contributed by Himalayan glaciers. This makes glacier meltwater a significant source of water in North and North-Eastern Indian rivers throughout the year (Bahadur 1988). The objective of this study is to understand the geo-chemical processes in the ice and meltwater to identify the main source of the major ionic components in Eastern Indian Himalayas.

2 Experimental Study

2.1 Study Area

Rathong glacier (Area-3.9 km^2; Agrawal and Tayal 2013) is located at the head of the Rathong valley between the latitude 27° 33′ N and 27° 35′ N, and longitude 88° 06′ E. Its, pro-glacial stream (Rathong chu) flows from 27° 33′ N and 88° 07′ E to 27° 17′ N and 88° 17′ E, through the Rathong valley and ultimately meets Tista river. Rathong valley is covered with transparent vegetation, forest cover, cultivated and agricultural land. However moderately sloped mountain sides are not vegetated and severely weathered. The study was carried out with the ice and meltwater samples from various locations from Rathong Glacier (Eastern Indian Himalayas) and it's pro-glacial stream Rathong chu. Sampling sites were located in the West district of Sikkim state of India.

2.2 Methodology

Ice and meltwater samples were collected using clean sampling techniques (Ferrari et al. 2000; Ostrem 1975) from East Rathong chu in pre-monsoon season, starting from snout at 4,678 m a.s.l. to downstream The samples were then immediately transferred into ultra-clean 250 ml Teflon bottles. Extreme care was taken during sample collection, handling, and storage to minimize the contamination. All the equipments used in the sampling and storage were cleaned as per described methods. All samples were kept frozen till the analysis. Measurements of pH, EC (electrical conductivity) and temperature of the ice and meltwater samples were done during the collection of samples by using portable pH and EC-meters (HI 9813-5, Hanna Instruments) and digital thermometer. Water samples were filtered through 0.45-μm Millipore membrane (Milex HA syringe driven filter unit) before injecting for Ion Chromatography.

The major cations (Li^+, Na^+, NH_4^+, K^+, Mg^{2+}, and Ca^{2+}) and anions (F^-, Cl^-, NO_2^-, SO_4^{2-}, NO_3^-, PO_4^{3-}, Br^-) were separated and quantified by using an ion-chromatography (Dionex ICS 2100 RFIC-ESP-EG™). The cations were separated by using Ion Pac® CS12 A column (20 mM methanesulfonic acid at 1.0 ml min^{-1} as eluent) with an Ion Pac® CG12 A precolumn, while separation of anions was performed on an Ion Pac® AS11-HC Analytical column (30 mM KOH at 1.5 mL min^{-1} as eluent) with an Ion Pac® AG11-HC precolumn. The instrument detection limits was 0.010 μeq L^{-1} for each ion. Linear dynamic range of concentrations for anions and cations was covered by different linear weighted calibration curves, one from 0.025 to 1 μeq L^{-1} and the others from 1 to 10 μeq L^{-1} and from 10 to 100 μeq L^{-1}.

3 Results and Discussion

3.1 Spatial Distribution and Concentration of Ions in Ice and Meltwater

The average pH of ice and meltwater samples was 8.13, suggesting mild alkaline nature, while EC ranged from 27 to 105 μS cm^{-1} (average 50.42 μS cm^{-1}) for meltwater samples. Statistical analysis of the results obtained from the chemical analysis of the samples shows that measured total cations (TZ^+) and total anions (TZ^-) are related by the equation $TZ^+ = 2.5116{*}TZ^- - 61.191$. Correlation coefficient between total cations and total anions is 0.692.

East Rathong chu has the dominance of Ca^{2+}, followed by Na^+ (17 %) > K^+ (7.8 %) > NH_4^+ (6.8 %) > Mg^{2+} (3.2 %), Li^+ (0.095 %) in the meltwater stream. There is marked increase in the concentrations of Ca^{2+}, K^+, Na^+, NH_4^+, Cl^-, NO_2^-, and NO_3^- between altitude 4,678 and 4,660 m a.s.l. This sharp rise in the concentration may be due the presence of huge amount of debris at the proglacial stream site of the glacier which is exposed to weathering (Collins 1979; Raiswell 1984; Sharp et al. 1995). Relatively low concentration of these cations and anions at the head of the proglacial stream Rathong chu may be because the river flows through the rocks of less reactive phases like schist, gneisses, granites and granodiorites of central crystalline zone; these would provide a little contribution to the solute load in the meltwater. There is a sharp increase in the concentrations of Na^+, NH_4^+, K^+, Mg^{2+} and NO_3^- at the lower stretch of Rathong chu, which is because the middle and lower reaches the river flows through comparably more reactive phases such as marbles, calcite and dolomite, which may result in relatively high ionic concentration.

Relatively very high concentration of SO_4^{2-} at the mouth of the proglacial stream may be because of the occurrence of pyritous-carbonaceous slates and phyllites in the geological units of higher Himalayas, which suggests that the primary source of high concentration of SO_4^{2-} near the source of the stream would be the oxidation of pyrites (Singh and Hasnain 2002). It was estimated that the contribution to the major ion chemistry in the mountainous catchments by atmospheric deposition of Na^+ and K^+ is up to 20 %, while for Ca^{2+}, Mg^{2+}, and SO_4^{2-} it is up to 5 %.

Scatter plot of ($Ca^{2+} + Mg^{2+}$) against the total cations (TZ^+) shows that all the points fall below the equiline (1:1 Line) (Fig. 1). Most of the points fall in the range of 0.60–0.80. Ratio of $(Ca^{2+} + Mg^{2+})/(Na^+ + K^+)$ is considerably high (4.05 %). These results indicate that weathering of carbonates is the major source

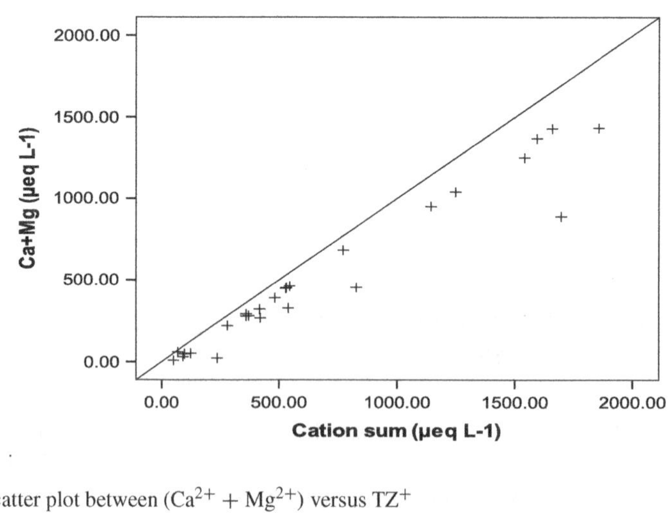

Fig. 1 Scatter plot between ($Ca^{2+} + Mg^{2+}$) versus TZ^+

of dissolves ions in the ice and meltwater of Rathong glacier (Pandey et al. 2001). Scatter plot of $[(Ca^{2+} + Mg^{2+})/(Na^{+} + K^{+})]$ with respect to altitude shows sharp increases at glacier and in the lower reaches of the proglacial stream Rathong chu. This confirms the presence of carbonate weathering at glacier because of debris and at lower reaches of the stream because of marbles, dolomite and calcite.

The ratio of cations (Na^{+} and K^{+}) to Cl^{-} can be used as indices owing to the abundant concentration of Cl^{-} in the ocean and its low level in most rocks (Stallard and Edmond 1983; Sarin et al. 1989). The average ratio of Na^{+} to Cl^{-} is 4.86 and K^{+} to Cl^{-} is 2 in ice and meltwater. These ratios are much higher than those expected from the marine aerosols. For marine aerosols expected values of Na^{+}/Cl^{-} and K^{+}/Cl^{-} ratio are 0.85 and 0.017, respectively (Pandey et al. 2001).

According to the study (Sun et al. 1998), evidences were found that post-depositional processes control the levels of ammonium ion in ice and meltwater. In present study, weak correlation of NH_4^{+} with SO_4^{2-} ($r^2 = -0.335$) and with NO_3^{-} ($r^2 = 0.339$) was found. The enrichment of NH_4^{+} relative to SO_4^{2-} can explain a post-depositional origin of NH_4^{+} (Marinoni et al. 2001). NH_4^{+}/SO_4^{2-} ratio has an average value of 2.56 and shows that only a fraction of NH_4^{+} in ice and meltwater is neutralized by SO_4^{2-}. The rest of NH_4^{+} is balanced by NO_3^{-} and NO_2^{-} as indicated by the average of the $NH_4^{+}/(SO_4^{2-} + NO_3^{-})$, which is 0.3. The low values of ratio $NH_4^{+}/(SO_4^{2-} + NO_3^{-})$, which is always <1 in our study also indicates that post-depositional scavenging of gaseous HNO_3 is the most active nitrate deposition mechanism in our study area. High proportion of NH_4^{+} in glacier ice and meltwater could be because of transportation of NH_3 from the low lying valleys and mountains to the glacier snow and ice. These low lying valleys have cultivated lands, vegetation and forests (Marinoni et al. 2001). Figure 2 indicates that the samples with lower ionic strength tend to have almost

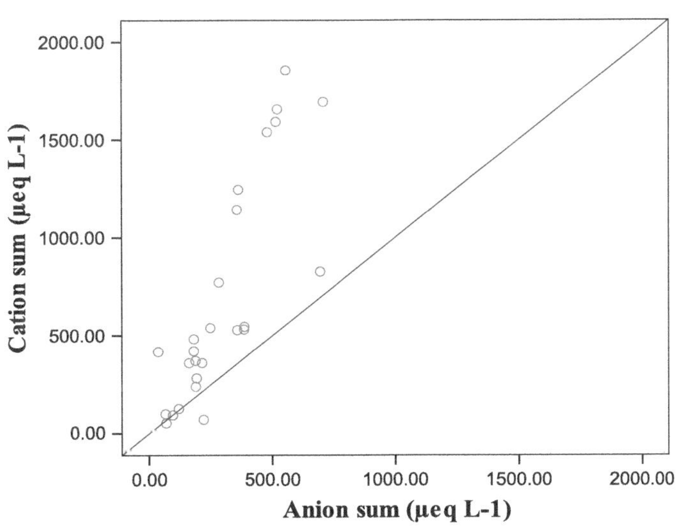

Fig. 2 Scatter plot between cation sum and anion sum

equal cation sum and anion sum, whereas increase in the ionic strength results in the high alkalinity in the samples and the sum goes above the equiline (1:1 line) (Krnavek et al. 2012).

3.2 *Factor Analysis*

In order to determine the main source of major ions, factor analysis on chemical data was performed (Fig. 3). For normal distribution of the variables, data was standardized to have a mean of zero and standard deviation of one (Singh and Hasnain 2002). In this study, correlation coefficients of Cl^- with K^+, Mg^{2+} and Ca^{2+} are 0.671, 0.214, and 0.401 respectively. This is very low compared to the values of correlation coefficients for marine water which is usually higher than 0.95 (Suzuki et al. 2002). These low values of correlation coefficients indicate that the sea salt may have negligible contribution to ionic concentration in ice and meltwater. Correlation between Ca^{2+} and $SO_4{}^{2-}$ is very weak ($r^2 = -0.036$), suggesting that Asian dust is not the major source for these particulate matters (Wake et al. 1993). Strong correlation between $NO_3{}^-$ and $NO_2{}^-$ ($r^2 = 0.858$) suggests the reduction of Nitrate into nitrite (Yang et al. 1995).

It was found that there are two statistically significant factors for the data set of our study, explained by 63.76 % of total variance. Factor-1, loaded with $NO_2{}^-$, Ca^{2+}, Na^+, Cl^-, $NH_4{}^-$, $NO_3{}^-$ and K^+ may be related to chemical weathering and post-depositional processes while factor-2 loaded with Mg^{2+}, $SO_4{}^{2-}$ and F^- may be originated from the dust storms (Singh and Hasnain 2002; Marinoni et al. 2001; Li et al. 2007; Negre and Roy 1998; Han et al. 2004; Kang et al. 1999).

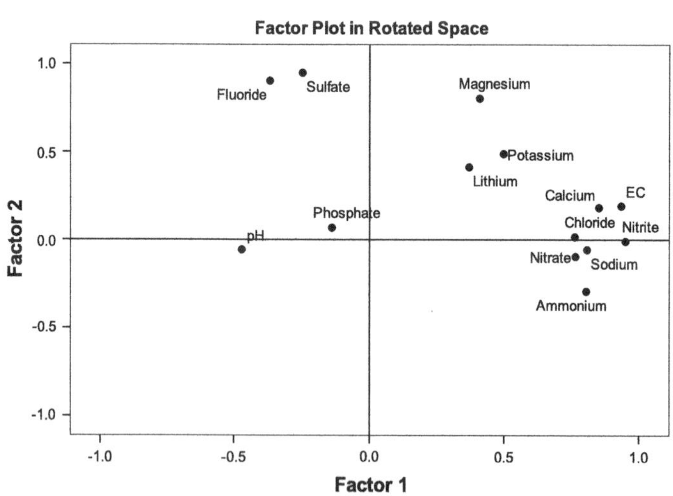

Fig. 3 Factor score plot of principal and oblimin rotated R-mode factor analysis

Thus factor analysis also supports our previous finding of this study that water chemistry is primarily controlled by the chemical weathering and post-depositional origin of major solutes. While secondary controlling factor is dust storm, which is controlling a lesser number of solutes.

4 Comparison with Chemical Composition of Meltwater from Western Himalayan Glaciers

Average chemical composition of meltwater from Rathong glacier (Eastern Himalayas) and other Western Himalayan glaciers has been summarized in Table 1. The abundance of Ca^{2+} was observed highest among all the cations from the glacier meltwaters and varied as Kafni > E. Rathong > Dokariani > Gangotr i > Dudu glacier. Among anions, Cl^- and NO_2^- are maximum in meltwater of E. Rathong glacier as compared to other glaciers in western Himalaya. Also, total ionic strength of meltwater seems to be significant in comparison to the size of different glaciers under study.

The average $(Ca^{2+} + Mg^{2+})/(Na^+ + K^+)$ ratios for Kafni, Dudu, Dokriani and Gangotri glaciers have been reported by different authors to be 7.8, 1.45, 2.06 and 2.55 (presented in Table 1) which indicate the dominance of carbonate weathering as a major source of dissolved ions in the glacier basin, even when calcite is present in low concentrations (Raiswell 1984). On the contrary, the sample taken from the East Rathong glacier snout indicate $(Ca^{2+} + Mg^{2+})/(Na^+ + K^+)$ ratio to be <1. $(Na^+ + K^+)/Cl^-$ ratios in this study also indicates varied patterns as compared to established trends for western Himalayas. However, East Rathong glacier indicates similarity to Gangotri and Dokriani glaciers with reference to high sulfate concentration being attributed to pyrite dissolution in the bed rock (Bhatt 1963).

Mean value of ionic ratios at E. Rathong chu, meltwater stream is significantly higher for Na^+/Cl^- and $(Na^+ + K^+)/TZ^+$ as compared to the ionic ratios reported by Singh and Hasnain (1999) for Bhagirathi and Alaknanda meltwater streams. Simultaneously, a lower value of $(Ca^{2+} + Mg^{2+})/TZ^+$ and $(Ca^{2+} + Mg^{2+})/(Na^+ + K^+)$ is also observed as compared to Bhagirathi and Alaknanda. This indicates a different set of weathering mechanism acting at East Rathong glacier valley which may be due to variation in geological setup of the bed rock. While this needs to be further validated through additional observations, geomorphological setup of proglacial stream consisting of at least three meltwater lakes downstream of glacier snout may also be significant regulator of ionic enrichment and variability of meltwater stream in comparison to western Himalaya streams.

Table 1 Chemical characteristics of ice and water samples from Rathong glacier and its proglacial stream Rathong chu (μeq L^{-1})

Ionic concentration	Na^+	K^+	Mg^{2+}	Ca^{2+}	Cl^-	NO_2^-	SO_4^{2-}	
E. Rathong	70	66	22	274	64	98	286	Present study
Kafni	65	31	165	587	35		76	Singh et al. (1998)
Dudu	43	25	7	91	5	15	85	Ahmad and Hasnain (2000)
Dokriani	64	116	100	271	16		406	Ahmad and Hasnain (2000)
Gangotri	75	83	197	206	11.2	1.9	401	Singh et al. (2012)

Ionic ratios	Na^+/Cl^-	$(Ca^{2+}+Mg^{2+})/TZ^+$	$(Na^++K^+)/TZ^+$	$(Ca^{2+}+Mg^{2+})/(Na^++K^+)$	SO_4^{2-}/Cl^-	K^+/Cl^-		
E. Rathong chu	8.90	0.66	0.29	3.23	21.31	3.42		Present study
Alaknanda	3.83	0.83	0.17	6.33	4.7	2.81		Singh and Hasnain (2002)
Bhagirathi	3.97	0.78	0.21	4.06	21.8	5.75		Singh and Hasnain (2002)

5 Conclusion

Chemical characteristics of East Rathong glacier meltwater and its proglacial stream E. Rathong chu validates the earlier conclusions related to solute acquisition in meltwater being proportional to residence time of meltwater in contact with bedrock/morainic sediments. Similarities were observed in comparison to meltwater streams from western Himalayas, but variability especially with reference to concentration of Na^+, K^+ and Cl^- ions needs to be further validated.

References

Agrawal A, Tayal S (2013) Assessment of volume change in East Rathong glacier, Eastern Himalaya. Int J Geoinformatics 9(1):73–82

Ahmad S, Hasnain SI (2000) Meltwater characteristics of Garhwal Himalayan glaciers. J Geol Soc India 56:431–439

Bahadur J (1988) Himalayan water from snow and glaciers. In: First national water conference proceedings, vol 2. Ministry of Water Resources, Government of India, pp 59–65

Bhatt BK (1963) Preliminary study of the Bhagirathi basin between Uttarkashi and Gomukh. In: Proceeding of national symposium on Himalayan geology, Calcutta, Geological Society of India, vol 15. Miscellaneous Publication, pp 1–8

Brown GH, Sharp MJ, Tranter M (1996) Subglacial chemical erosion: seasonal variations in solute provenance, Haut Glacie d' Arolla, Valais, Switzerland. Ann Glaciol 22:25–31

Brown GH, Sharp MJ, Tranter M, Gurnell AM, Nenow PW (1994) Impact of post-mixing chemical reactions on the major ion chemistry of bulk meltwaters draining the Haut Glacie d' Arolla, Valais, Switzerland. Hydrol Process 8:465–480

Collins DN (1979) Hydrochemistry of meltwater draining from an Alpine glacier. Arct Antarct Alp Res 11:307–324

Collins DN, Hasnain SI (1995) Runoff and sediment transport from glacierised basin at the Himalayan scale. In: Symposium on interpretation and management of sediment and water quality, boulder, Colorado, vol 226. International Association of Hydrological Sciences Publication, pp 17–25

Drewry D (1986) Glacial geologic processes. Edward Arnold Publication Pvt. Ltd., London

Ferrari CP, Moreau AL, Boutron CF (2000) Clean conditions for the determination of ultra-low levels of mercury in ice and snow samples. Fresenius' J Anal Chem 366:433–437

Han YX, Xi XX, Song LC, Ye YH, Li YH (2004) Spatiotemporal sand-dust distribution in Qinghai-Tibet plateau and its climatic significance. J Desert Res 25:588–592

Hasnain SI, Subramanian V, Dhanpal K (1989) Chemical characteristics and suspended sediments load of meltwater from a Himalayan glacier in India. J Hydrol 106:99–106

Kang SC, Qin DH, Yao TD, Wake CP, Mayewski PA (1999) Summer monsoon and dust signals recorded in the Dasuopu firn core, central Himalaya. Chin Sci Bull 44:2010–2015

Krnavek L, Simpson WR, Carlson D, Domine F, Douglas TA, Sturm M (2012) The chemical composition of surface snow in the Arctic: examining marine, terrestrial, and atmospheric influences. Atmos Environ 50:349–359

Li CL, Kang SC, Zhang QG, Kaspari S (2007) Major ionic composition of precipitation in the Nam Co region, Central Tibetan Plateau. Atmos Res 85:351 360

Marinoni A, Polesello S, Smiraglia C, Valsecchi S (2001) Chemical composition of freshsnow samples from the southern slope of Mt. Everest region (Khumbu-Himal region, Nepal). Atmos Environ 35:3183–3190

Negre LP, Roy S (1998) Chemistry of rainwater in the Massif Central (France): a strontium isotope and major element study. Appl Geochem 13:941–952

Ostrem G (1975) Sediment transport in glacial meltwater streams. In: Jolping AV, Mc Donald (eds) Glacio-fluvial and Glacio-lacustrine sedimentation. Society economic palaeont and mineral, vol 23. Special Publication, pp 101–122

Pandey SK, Singh AK, Hasnain SI (2001) Hydrochemical characteristics of meltwater draining from Pindari glacier, Kumaon Himalaya. J Geol Soc India 57:519–527

Puri VMK (1994) Elements of glaciology. Geol Surv India 79:215–249

Raiswell R (1984) Chemical models of solute acquisition in glacial meltwater. J Glaciol 30:49–57

Sarin MM, Krishnaswamy S, Dilli K, Somayajulu BLK, Moore WS (1989) Major ion chemistry of Ganga-Brahmaputra river system: weathering processes and fluxes of Bay of Bengal. Geochim Cosmochim Acta 53:997–1009

Sharp M, Brown GH, Tranter M, Willis IC, Hubbard B (1995) Comments on the use of chemically based mixing models in glacier hydrology. J Glaciol 41:241–246

Singh AK, Hasnain SI (1999) Major ion chemistry and weathering control in a high altitude basin: Alaknanda river, Garwal Himalaya, India. Hydrol Sci J 43:825–843

Singh AK, Hasnain SI (2002) Aspects of weathering and solute acquisition processes controlling chemistry of sub-Alpine proglacial streams of Garhwal Himalaya, India. Hydrol Process 16:1–15

Singh AK, Pandey SK, Panda S (1998) Dissolved and sediment load characteristics of Kafni glacier meltwater, Pindar valley, Kumaon Himalaya. J Geol Soc India 52:305–312

Singh VB, Ramanathan AL, Pottakkal JG, Sharma P, Linda A, Azam Md F, Chatterjee C (2012) Chemical characterization of meltwater draining from Gangotri Glacier, Garhwal Himalaya, India. J Earth Syst Sci 121(3):625–636

Stallard RF, Edmond JM (1983) Geochemistry of the amazon: the influence of the geology and weathering environment on the dissolved load. J Geophys Res 88:9671–9688

Sun JY, Qin DH, Mayewski PA, Dibb JE, Whitlow S, Li ZQ, Yang QZ (1998) Soluble species in aerosol and snow and their relationship at glacier 1, Tien Shan, China. J Geophys Res-Atmos 103(D21):28021–28028

Suzuki T, Iizuka Y, Matsuoka K, Furukawa T, Kamiyama K (2002) Distribution of sea salt components in snow cover along the traverse route from the coast to Dome Fuji station 1000 km inland at east Dronning Maud Land, Antarctica. Tellus 55B:407–411

Tranter M, Brown G, Raiswell R, Sharp M, Gurnell A (1993) A conceptual model of solute acquisition by Alpine glacier meltwaters. J Glaciol 39:573–581

Wagenbach D (1989) Environmental records in alpine glaciers. In: Oeschger H and Langway CC (eds) The environmental record in glaciers and ice sheets, Dahlem, Konferenzen. Wiley and sons, Chichester, pp 69–83

Wake CP, Mayewski PA, Zichu X, Ping W, Zhonggin L (1993) Regional distribution of monsoon and desert dust signals recording in Asian glaciers. Geophys Res Lett 20:1411–1414

Yang Q, Mayewski PA, Whitlow S, Twickle RM, Morrison M, Talbot R, Dibb J, Linder E (1995) Global perspective of nitrate flux in ice cores. J Geophys Res, 100(D3):5113–5121

Socio-economic Dimension of Snow and Glacier Melt in the Nepal Himalayas

Narayan P. Chaulagain

Abstract A mass balance analysis of the glaciers in the Nepal Himalayas has revealed that there have been accelerated retreats of glaciers causing widespread adverse impacts on the ecosystem, environment and economy in Nepal. The decreasing water storage in the mountain due to snow and glacier melt has created additional threats to already fragile and vulnerable mountain communities. The people and ecosystem in the mountain and foot hills are already under immense pressure from growing economic development. Climate change is putting additional pressure on the lives and livelihoods of the mountain people and on the ecosystem they depend upon. Studies have revealed that the majority of the glaciers in Nepal Himalayas are already retreating so rapidly that even without any further warming; most of them will disappear by the end of this century. This may result in decreased melt-water contribution to total water availability, particularly during dry seasons. The hydropower and irrigation sector, which are already under stress during non-monsoon seasons, will be badly affected. Though the annual decrease in per capita water availability due to warming may not be very much significant, the seasonal effect of the warming particularly during non-monsoon season is substantial. This will have major adverse impact to economy and society of Nepal. Particularly, the poor, marginalized, subsistence farmers and those living on natural system will face the largest adverse impact.

Keywords Glacier • Glacial lake outburst floods • Climate change • Socio-economy • Water resources

N.P. Chaulagain (✉)
ADAPT Nepal, Kathmandu, Nepal
e-mail: narayanchaulagain@gmail.com

R. Joshi et al. (eds.), *Dynamics of Climate Change and Water Resources of Northwestern Himalaya*, Society of Earth Scientists Series,
DOI 10.1007/978-3-319-13743-8_15

1 Introduction

Nepal has more than 6,000 rivers flowing from mountains to hills and plains. There are about 3,808 glaciers with a total area of 4,212 km^2 in Nepal Himalayas (ICIMOD 2011), which provide perennial flows for major river systems in Nepal. The contribution of snow and glacier melt water in these rivers, particularly during dry season is substantial. The temperature in Nepal is increasing faster in higher altitudes than in the lower ones, resulting in accelerated melting of glaciers, formation of glacial lakes in the mountain valleys and expansion of existing glacial lakes in Nepal Himalayas. This has resulted in increased risks of glacial lake outburst floods creating threats to people, property, infrastructures, livestock and ecosystem not only in the mountains but also far downstream in the hills and plains. Studies have suggested that glacial lake outburst floods in the Himalayas have occurred more frequently during the last 50 years (Mool et al. 2010).

Accelerated melting of snow and glaciers in the Himalayas has adversely affected the water storage capacity of the mountains. Decreased melt water contribution to the river flows particularly during non-monsoon season has negative impact on run-of-river hydropower, irrigation and even municipal water supply when the demand of water is relatively higher. The impacts of climate change are disproportionately distributed within the country among different communities and sectors of society. The poorer, marginalised, people of the high mountains, subsistence farmers are likely to suffer the earliest and the most (Jianchu et al. 2007). Mountain people have lived with and survived great hazards for thousands of years, but current rates of climate change have made them unable to cope with the changes by imposing severe and uncertain socioeconomic pressures.

2 Impacts on Water Availability

Increased temperature will not only affect the annual glacier mass balance, but also will change precipitation pattern, i.e. more rainfall and less snowfall. A study in the Langtang valley with a catchment area of 340 km^2 in central Nepal has shown that 1 °C rise in average temperature may reduce snow-to-rain ratio from 1.6 to 1.2 (Chaulagain 2009). Rainfall, unlike snowfall, will not be stored in the mountains, but will immediately be drained out from the basin resulting in less ground water recharge upstream and more floods downstream during the monsoon.

Majority of the glaciers in Nepal Himalayas are retreating at the rate higher than the glaciers elsewhere in the world, though the rate of retreat varies from glacier to glaciers. A sensitivity analysis of all the glaciers in Nepal Himalayas was done by using glacier mass balance model developed by Y. Ageta (Kadota and Ageta 1992; Naito et al. 2001) for the glacier AX010 in the eastern Himalayas and by applying the same empirical equation for the 24 glaciers in the Langtang Himalayas of the central Nepal. The analysis has shown that many of glaciers in

Nepal Himalayas will disappear by the end of this century in the Nepal Himalayas, if the melting of the glaciers will continue at the present rate. Likewise, the analysis has revealed that the current ice reserve of 480.6 km^3 in Nepal Himalayas will come down to 0.6 km^3 by 2,100, if the temperature will increase by 0.03 °C per year (see Fig. 1).

Such an accelerated decrease of ice reserve in Nepal Himalayas may result in the change of melt water contribution to the annual river flows. Currently, annual melt-water from the glaciers in the Nepal Himalayas is about 6 km^3 (see Fig. 2).

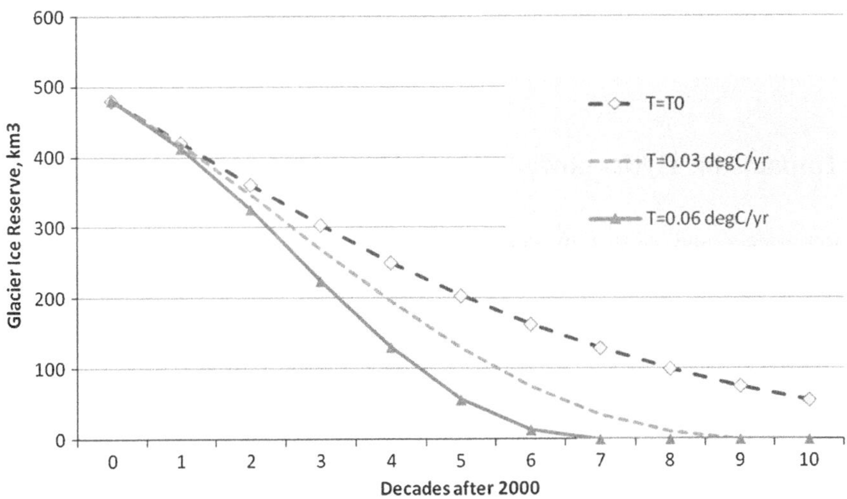

Fig. 1 Sensitivity of glacier ice reserve to warming. *Source* Modified after Chaulagain (2007)

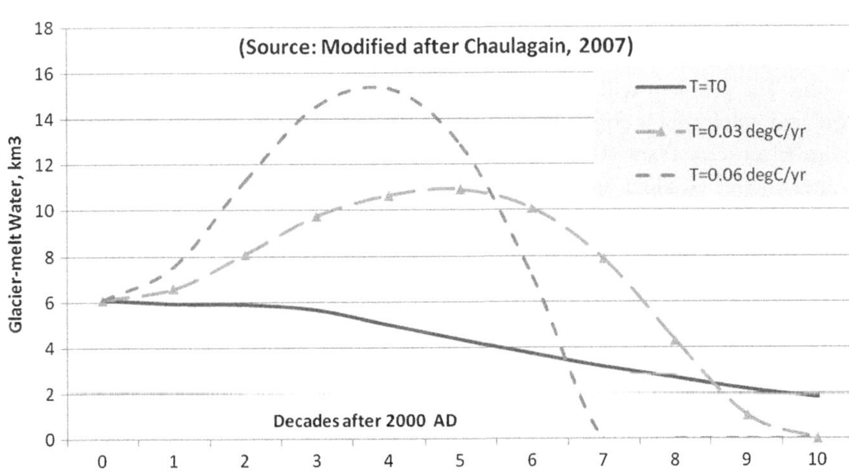

Fig. 2 Sensitivity of glacier-melt water to warming. *Source* Modified after Chaulagain (2007)

Water supplies stored in glaciers and snow cover are projected to decline in the course of the century, thus reducing water availability during warm and dry periods (Bates et al. 2008). The glacier-melt water contribution will initially rise with the rise in temperature and ultimately will go down after the contributing glaciers disappear (IPCC 2007). Decreasing glacier-melt water and disappearance of glaciers will ultimately change the hydrograph of the river system. The snow- and glacier-fed rivers will be converted into rain-fed ones. The monsoon streamflows including flood water will further increase and the dry season streamflows will further decrease. Currently, the snow and glacier-melt contributes to about 13 % of Nepal's total annual surface water. However, the snow and glacier-melt contribution is significant (up to 34 %) during April–May, when the demand for water is relatively higher.

3 Impacts on Hydropower Generation

Flood waters and river flows during monsoon seasons have a relatively less meaning for hydropower generation in the Nepalese context where majority of the hydropower plants are of the run-of-river type. As the glacier-melt water initially increases with increase in temperature and hydropower potential generally may increase as it largely depends on the lean season flows. Acharya (2011) after analyzing the observed flows of 6 major tributaries of Narayani River (i.e. Marsyangdi, BudhiGandaki, Trisuli, Kali Gandaki, Madi and Seti) for the period of 20–40 years has reported that there was a decreasing trend of the average flow during dry season (November–April). Nepal's hydropower generation generally follows the pattern of dry season flows (Chaulagain 2007). Over 90 % of Nepal's existing hydropower plants are the runoff river type which are generally designed based on the dry season flows.

These power plants have been already facing the problem of water shortages during dry seasons and generating only about 30 % of the total installed capacity in dry months. The problem will be further exacerbated during dry season by the reduced snow and glacier-melt contribution in the future. A similar analysis of Marsyangdi Hydro Electricity Plant (HEP) (69 MW) by using the outputs of the glacier mass balance model (Kadota and Ageta 1992; Naito et al. 2001) has revealed that out of the existing average annual electricity generation, about 27 % is contributed by snow and glacier melt water. Further increase in temperature will adversely affect the projected future energy generation of the hydropower station. The result of the sensitivity analysis of annual electricity generation is given in the Fig. 3.

As given in the Fig. 3, the future impact on the electricity generation by the hydropower plants largely depends on the rate of temperature increase. At the initial decades (e.g. till 2030), the electricity generation potential increase with the rate of temperature rise and then gradually goes down. The higher the rate of temperature increase in the future, the earlier the peak of the electricity generation potential of the hydropower plant will arrive.

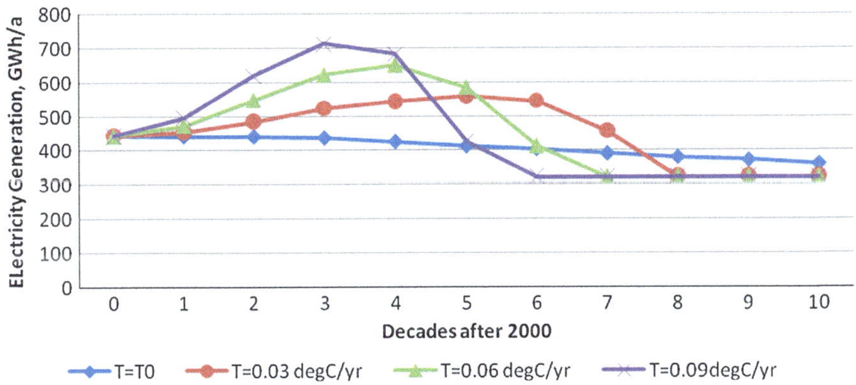

Fig. 3 Sensitivity of annual electricity generation of Marsyangdi HEP (2001–2100 AD)

4 Impacts on Irrigation and Agriculture

Increased temperature will result in increased evapo-transpiration leading to increased irrigation water demand, and decreased river flows. Studies have suggested that glacier-fed perennial rivers will be converted into rainfed seasonal rivers after the glaciers disappear. The ratio of maximum to minimum flows of rain-fed rivers is substantially higher than that of snow-fed rivers indicating a possible future re-distribution of water among months after glacier-fed rivers will become the rain-fed ones. Furthermore, increased temperatures will widen the gaps between the water supply and demand for irrigation. A water balance analysis of Bagmati river basin at Kathmandu Valley has shown that a one-degree rise in annual temperature may increase the water demand by 3.7 % and reduce the annual river flow by 1.5 % simultaneously, while a three-degree rise in annual temperature may increase the water demand by 11.2 % and reduce the annual river flow by 4.4 % (see Fig. 4).

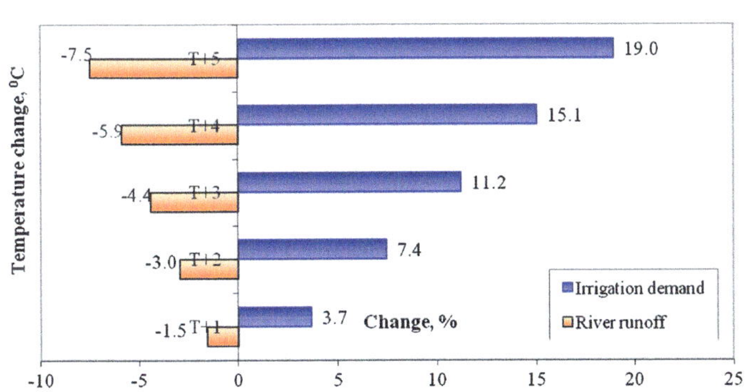

Fig. 4 Sensitivity of annual water balance situation at Kathmandu valley at chovar. *Source* Chaulagain (2007)

Due to monsoon dominated flow pattern, the Kathmandu valley in Bagmati river basin has already been facing water shortage during non-monsoon seasons though there is surplus water during monsoon. The increase in temperature will further worsen the situation of too much water during rainy season and too little of it during dry season. Increases in frequency of climate extremes may lower crop yields (Tubiello et al. 2007). Increased evapo-transpiration and increased soil moisture deficit due to increased temperature may have significant adverse impacts on agriculture production and food security. Water availability is a key component of food security as the availability of water supplies is the single most important factor in food production (McGuigan et al. 2002). Changes in glacier melt, along with other changes in high-altitude hydrology, will affect agricultural production (Malone 2010). The small landholdings, subsistence farmers and the poorest of the poor will face the biggest adverse impact of reduced agriculture production due to reduced water availability, which ultimately may lead to the famine.

5 Impacts on Extreme Events

The shrinkage and disappearance of mountain glaciers may result in changes in the flow characteristics of glacier fed rivers and changes in flood severity and frequency (Kaltenborn et al. 2010). Glaciers in Nepal Himalayas have been thinning and retreating at rates of 10–60 m per year and many small glaciers with surface area of less than 0.2 km^2 have already disappeared (Bajracharya et al. 2007). Upward shifts in the elevation of a terminus as great as 100 m have been recorded during the past 50 years and retreat rates of 30 m per year are common in Nepal Himalayas (Malone 2010). Increased melt of snow and glaciers in Nepal Himalayas has resulted in formation of glacial lakes and expansion of existing ones in the mountain valleys (Ives et al. 2010). Higher temperature may increase the likelihood of precipitation falling as rain rather than snow (IPCC 2007), which may result in increased likelihoods of floods during rainy season and decreased river flows during dry season. Chaulagain (2009) has revealed that decrease in snow cover areas has exponentially increased the ratio of maximum-to-minimum stream-flows (i.e. increased maximum flows and decreased minimum flows simultaneously) in Nepalese rivers. Moreover, increased melting of snow and ice including permafrost has induced an erodible state in the mountain soil which was previously non-erodible. This has increased likelihoods of landslides in the mountains. Because of warming, snowmelt begins earlier and winter becomes shorter, which affects river regimes, natural hazards, water supplies, infrastructures and people's livelihoods (Jianchu et al. 2007).

6 Socio-economic Consequences

Increase in evaporation, reduction in snow cover, and fluctuations in precipitation are key factors contributing to the degradation of mountain ecosystems. While too little water leads to vulnerability of production, too much water can also have adverse effects on crop productivity. Heavy precipitation events, excessive soil moisture and flooding disrupt food production and rural livelihoods (Bates et al. 2008). Changes in glacier regimes and runoff from snow and ice, combined with changes in precipitation timing and intensity may increase human vulnerability and may affect agriculture, forestry, health conditions and tourism (Kaltenborn et al. 2010). Decreased snow cover due to warming may result in direct adverse impacts on tourism. Rayamajhi (2012) has revealed that absence of snow on mountain caps may degrade the aesthetic view of the mountain and divert the tourists to other destinations. Increased risks of avalanches and glacier lake outburst floods in the mountain due to accelerated melting of snow and glaciers may adversely affect the tourist arrival in the mountains.

Water infrastructure, usage patterns and institutions have been developed in the context of current and past climatic conditions. Any substantial change in the frequency of floods and droughts, or in the quantity and quality or seasonal timing of water availability, will require adjustments that may be costly, not only in monetary terms but also in terms of societal and ecological impacts, including the need to manage potential conflicts between different interest groups. Increased risk of food and water shortage, water shortages for settlements, industry and societies; reduced hydropower generation potentials; potential for population migration due to floods and landslides are some of likely major adverse impacts associated with water resources.

Enhanced melting and increased length of the melt season of glaciers leads at first to increased river runoff and discharge peaks, while in the longer time-frame, glacier runoff is expected to decrease. The demand for water use is generally driven by changes in population, food consumption, economy, water pricing, technology, lifestyle and societal views regarding the value of freshwater ecosystems (Parry et al. 2007). The future socio-economic pathways will most likely increase the future water demand resulting in widening gap between water supply and demand, which will further exacerbate the existing water stress particularly during dry season. Currently, Nepal's annual renewable water availability is 7,656 m^3 per person, which is well above the global average water availability of 6,000 m^3 per capita per year and the water stress level of 1,700 m^3 per capita per year (FAO 2007). The analysis of different scenarios of future temperature increase and the United Nations population projection have revealed that the annual renewable water availability in Nepal even in 2100 AD will be above the critical line of water stress (see Fig. 5).

As the population continues to increase, more people will require more water for the cultivation of food, fibre and industrial as well as for livestock. Despite a significant surplus in annual water availability, many parts of Nepal, particularly urban and semi-urban areas are already facing seasonal water scarcity particularly during dry seasons. The seasonal water scarcity will most likely be more pronounced under the climate change. Availability of water at right time and right place is very much crucial for meeting the water demands. Imbalances between availability and demand, the degradation of water quality, inter-sectoral competition, and interregional and international conflicts, all bring water issues to the fore front (FAO 2007). The productivity of agricultural, forestry and fisheries systems depends critically on the temporal and spatial distribution of water (Bates et al. 2008).

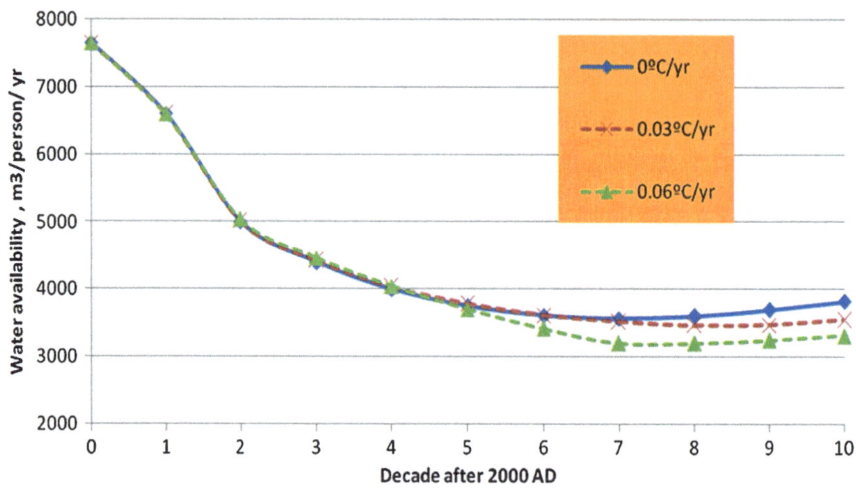

Fig. 5 Sensitivity of per capita water availability to temperature rise (2000–2100)

The impact of water scarcity is unevenly distributed among the sectors and income levels. Water scarcity is an issue of poverty. Unclean water and lack of sanitation are the major water issue for poor people. Water scarcity for poor people is not only about droughts or rivers running dry but it is also about guaranteeing the fair and safe access they need to sustain their lives and livelihoods (FAO 2007). Decreased runoff as a result of climate change will make it harder to improve safe access to drinking water, which leads to additional costs for the water supply sector and higher socio-economic impacts and follow-up costs. In the areas where extreme events become more intense and more frequent with climate change, the socio-economic costs of those events will increase significantly. Poor communities can be particularly vulnerable in such areas.

7 Conclusion

The melting of snow and glaciers and subsequent changes in water system have multi-facet impacts on society and economy because of direct linkage of water with people, ecosystem, economy and society. The impact of changed runoff regime and widening gaps between water supply and demand disproportionately falls more on poor, marginalized, subsistent farmers and the economic units which are directly dependent on natural system.

References

Acharya S (2011) Changing climatic parameters and its impact assessment in hydropower generation in Nepal: a case study on Gandaki river basin. In: Masters thesis submitted to the Department of Mechanical Engineering. Tribhuwan University, Kathmandu, Nepal

Bajracharya SR, Mool PK, Shrestha AB (2007) Impact of climate change on Himalayan glaciers and glacial lakes: case studies on GLOF and associated hazards in Nepal and Bhutan. In: Technical report of International Centre For Integrated Mountain Development, Kathmandu, p 119

Bates BC, Kundzewicz ZW, Wu S, Palutikof JP (2008) Climate change and water. In: Technical paper of the intergovernmental panel on climate change, IPCC Secretariat, Geneva

Chaulagain NP (2007) Impacts of climate change on water resources of Nepal: the physical and socio-economic dimensions. Shaker Verlag, Achen

Chaulagain NP (2009) Climate change impacts on water resources of Nepal with reference to the glaciers in the Langtang Himalayas. J Hydrol Meteorol 6(1):58–65

FAO (2007) Coping with water scarcity: challenge of the twenty-first century. UN-Water, FAO, Rome, p 29

ICIMOD (2011) Glacial lakes and glacial lake outburst floods in Nepal. In: Technical report of International Centre for Integrated Mountain Development, Kathmandu, p 9

IPCC, Climate Change (2007) Impacts, adaptation and vulnerability. In: Parry ML, Canziani OF, Palutikof JP, van der Linden PJ, Hanson CE (eds) Contribution of working group II to the fourth assessment report of the intergovernmental panel on climate change. Cambridge University Press, Cambridge, p 976

Ives JD, Shrestha RB, Mool PK (2010) Formation of glacial lakes in the Hindu Kush-Himalayas and GLOF risk assessment. In: Technical report of International Centre for Integrated Mountain Development, Kathmandu, p 6

Jianchu X, Shrestha A, Vaidya R, Eriksson M, Hewitt K (2007) The melting Himalayas: regional challenges and local impacts of climate change on mountain ecosystems and livelihoods. In: Technical paper of International Centre for Integrated Mountain Development, Kathmandu, p 15

Kadota T, Y. Ageta (1992) On the relation between climate and retreat of glacier AX010 in the Nepal Himalaya from 1978 to 1989. In: Bulletin of glacier research 10. Data centre for glacier research, Japanese Society of Snow and Ice, pp 1–10

Kaltenborn BP, Nellemann C, Vistnes II (2010) High mountain glaciers and climate change—challenges to human livelihoods and adaptation. United Nations Environment Programme, GRID-Arendal, Arendal

Malone E (2010) Changing glaciers and hydrology in Asia: addressing vulnerabilities to glacier melt impacts. USAID, Washington, DC, p 113

McGuigan C, Reynolds R, Wiedmar D (2002) Poverty and climate change: assessing impacts in developing countries and the initiatives of the international community. London School of Economics, London 39

Mool PK, Shrestha R, Ives J (2010) Glacial lakes and associated floods in the Hindu Kush-Himalayas. Information sheet #2/10, International Centre for Integrated Mountain Development, Kathmandu, p 4

Naito N, Ageta Y, Nakawo M, Waddington ED, Raymond CF, Conny H (2001) Response sensitivities of a summer-accumulation type glacier to climate change indicated with a glacier fluctuation model. In: Bulletin of glaciological research 18, Japanese Society of Snow and Ice, pp 315–322

Parry ML, Canziani OF, Palutikof JP, van der Linden PJ, Hanson CE (eds) (2007) Climate change 2007: impacts, adaptation and vulnerability. Contribution of Working Group II to the Fourth Assessment Report of the Intergovernmental Panel on Climate Change, Cambridge University Press, Cambridge, UK

Rayamajhi S (2012) Linkage between tourism and climate change: a study of the perceptions of stakeholders along the Annapurna trekking trail. In: Nepal tourism and development review 2. ISSN: 2091-2234

Tubiello FN, Soussana JF, Howden SM (2007) Crop and pasture response to climate change. Proc National Acad Sci 104(50):19686–19690

Author Index

A
Adhikari, V., 129
Anthwal, A., 115

B
Banerjee, Argha, 57
Bharat, Girija K., 181
Bhutiyani, M.R., 85

C
Chakraborty, Parthasarathi, 181
Chulagain, N.P., 191

D
Dobhal, D.P., 141

G
Goswami, Anuj, 67

J
Jain, Sanjay K., 35, 45
Joshi, Rajesh, 1, 9, 67
Joshi, S., 129
Joshi, S.C., 115
Joshi, Varun, 115

K
Khan, Abul Amir, 68
Kumar, Kireet, 130
Kumar, Rakesh, 130
Kuniyal, J.C., 98

L
Lal, Ravish, 67
Lohani, A.K., 35, 45

P
Palni, L.M.S., 1, 9
Pande, T., 129
Pandit, J., 1
Pant, Charu C., 151
Pant, Naresh C., 67
Pratap, Bhanu, 141

R
Rawat, Pradeep K., 151

S
Shankar, R., 57
Sharma, Brij Mohan, 181
Sharma, H., 129
Shrestha, Kedar Lal, 23
Singh, R.D., 35, 45

T
Tayal, Shresth, 181

R. Joshi et al. (eds.), *Dynamics of Climate Change and Water Resources of Northwestern Himalaya*, Society of Earth Scientists Series,
DOI 10.1007/978-3-319-13743-8

Subject Index

A

Aerosol, 4, 86, 97, 98, 101, 102, 104, 105, 107–110, 185
Aerosol optical depth (AOD), 97–99

B

Black carbon (BC), 10, 97–99, 104, 105
Brahmaputra, 1, 4, 23, 67–69, 71, 76, 79, 80, 82, 83, 142

C

Carbon dioxide (CO_2), 115–117, 119, 120, 123
Chemical characterization, 5
Chemical composition, 5, 100, 181, 187
Chorabari glaciers, 5, 141, 146, 148
Climate, 3, 19, 23, 25, 28, 29, 31, 36, 40, 46, 57, 85–87, 94, 98, 110, 115, 133, 141, 144, 149, 159, 160, 174, 182, 192, 196
Climate change, 1–4, 24–26, 85, 86, 93, 99, 116, 151, 162, 174, 191
Climate warming, 2, 40, 85, 116
Compound glaciers, 129, 137

D

Dabka catchment, 5, 152, 163, 174
Debris covered glacier, 57, 59, 61
Deforestation, 101, 159, 174, 175
Discharge ratios, 129, 136, 137
Diurnal, 4, 89, 92, 104, 119, 123, 130
Dokriani glacier, 3, 57, 59, 62, 143, 146, 187
Dunagiri glacier, 5, 141, 146

F

Flow duration curves, 129, 132, 134, 135
Flowline models, 57, 59
Fluvial deposit, 5, 151, 167

G

Ganges, 4, 11, 23, 29, 67, 68, 70, 73, 74, 76–78, 83
Gangotri, 4, 11, 58, 131, 132, 134, 144
Geohydrological processes, 151
Geological Survey of India, 143
Glacial lake, 3, 35–39, 41, 45–49, 51, 54
Glacier lake outburst floods (GLOF), 36, 45, 197
Glacier modeling, 57, 59
Glaciers, 1, 2, 4, 11, 23, 24, 31, 36, 37, 42, 46, 58, 86, 130, 141, 149, 182, 187, 191, 192, 194, 198
GLOF modeling, 45, 54

H

Hazards, 35, 39, 160, 175, 176, 192
Himalaya, 1, 2, 5, 9, 23, 36, 37, 45, 58, 68, 86, 88, 94, 98, 115, 130, 141, 142, 148, 151, 174, 181, 187, 191
Himalayan Rivers, 1, 23, 29, 130, 132
Hunza basin, 3, 28, 31

I

Indian Himalaya, 4, 9, 98, 142
Indian Himalayan Region (IHR), 98

R. Joshi et al. (eds.), *Dynamics of Climate Change and Water Resources of Northwestern Himalaya*, Society of Earth Scientists Series,
DOI 10.1007/978-3-319-13743-8

Indus, 1, 4, 23, 67, 69, 71, 72, 74, 83, 142, 182
International Centre for Integrated Mountain Development (ICIMOD), 46, 192
Ions, 5, 181, 183, 186
IPCC, 2, 24, 26, 28, 36, 86, 98, 196

K
Kosi basin, 4, 88, 131, 132, 151, 152
Kumaun Himalaya, 4, 155

L
Lesser Himalayas, 47, 131, 152, 155
Livelihood, 11, 23, 170, 197, 198

M
Mass balance, 5, 29, 37, 58, 60, 142, 146, 148, 149, 191, 194
Meltwater, 5, 130, 132, 181–183, 185, 187, 189
MIKE 11, 3, 53, 54
MODIS, 2, 9, 11, 12, 14, 20, 106
Monsoon, 1, 67, 83, 85, 89, 107, 129, 144, 175, 183, 192, 194

N
NDSI, 9, 14
Nepal Himalaya, 88, 191, 192, 196
Non-glacial river basins, 5, 151
Non-perennial springs, 151, 164, 174
Northwestern Indian Himalaya, 99, 110

P
Perennial springs, 151, 164, 177
Phenology, 115
Pollution sources, 99
Precipitation, 2, 27, 29, 67, 70–72, 74, 76, 80, 82, 83, 85, 87, 90, 129, 169, 174
Pro-glacial stream, 5, 181, 182

R
Rathong glacier, 5, 181, 182, 185, 189
RCMs, 25, 26
Receding, 142, 144
Remote sensing, 9, 10, 25, 42, 48, 70, 154
Retreat, 5, 35–37, 39, 61, 62, 141, 145, 191
Runoff, 3, 12, 17, 24, 29, 31, 130, 134, 164, 182, 194, 198

S
Scenario, 2, 24, 31, 32, 141, 197
Seasonal snow cover, 10, 12, 19, 20
Simulation, 3, 29, 42, 45, 49, 50
Simulation of GLOF, 45
Snow and Glacial Melt Runoff, 29
Snow cover monitoring, 2
Snow melt, 29, 31, 32, 142
Socio-economic, 11, 87, 196, 198
Spring hydrology, 5, 151, 154, 164, 170, 174
Spring hydrology of non-glacial, 151
Streamflow, 12, 194
Sustainable, 2, 24, 131, 151, 175

T
Temperature, 2, 11, 24, 26, 28, 36, 85, 88, 97, 105, 107, 117, 162, 195
Tipra glacier, 5, 141, 144, 145
Tourism, 2, 138, 197
TRMM, 4, 67, 70, 71, 74, 76, 79, 82

U
Upper Bhagirathi Basin, 11–13, 18
Upper Kosi basin, 4, 132
Uttarakhand, 1, 2, 4, 11, 46, 57, 115, 131, 142, 143

W
Warming, 20, 26, 36, 46, 58, 85, 91, 98, 191, 197
Water resources, 2, 17, 23, 24, 36, 46, 133
Western Himalaya, 4, 10, 88, 142, 181, 187